EGYPTIAN MYTH Geraldine Pinc
EIGHTEENTH-CENTURY BRITAIN Paul Langford
THE ELEMENTS Philip Ball
EMOTION Dylan Evans
EMPIRE Stephen Howe
ENGELS Terrell Carver
ETHICS Simon Blackburn
THE EUROPEAN UNION John Pinder
EVOLUTION Brian and Deborah Charlesworth
EXISTENTIALISM Thomas Flynn
FASCISM Kevin Passmore
FEMINISM Margaret Walters
THE FIRST WORLD WAR Michael Howard
FOSSILS Keith Thomson
FOUCAULT Gary Gutting
THE FRENCH REVOLUTION William Doyle
FREE WILL Thomas Pink
FREUD Anthony Storr
FUNDAMENTALISM Malise Ruthven
GALILEO Stillman Drake
GANDHI Bhikhu Parekh
GLOBAL CATASTROPHES Bill McGuire
GLOBALIZATION Manfred Steger
GLOBAL WARMING Mark Maslin
HABERMAS James Gordon Finlayson
HEGEL Peter Singer
HEIDEGGER Michael Inwood
HIEROGLYPHS Penelope Wilson
HINDUISM Kim Knott
HISTORY John H. Arnold
HOBBES Richard Tuck
HUMAN EVOLUTION Bernard Wood
HUME A. J. Ayer
IDEOLOGY Michael Freeden
…OSOPHY
JUDAISM
JUNG Anthony Stevens
KAFKA Ritchie Robertson
KANT Roger Scruton
KIERKEGAARD Patrick Gardiner
THE KORAN Michael Cook
LINGUISTICS Peter Matthews
LITERARY THEORY Jonathan Culler
LOCKE John Dunn
LOGIC Graham Priest
MACHIAVELLI Quentin Skinner
THE MARQUIS DE SADE John Phillips
MARX Peter Singer
MATHEMATICS Timothy Gowers
MEDICAL ETHICS Tony Hope
MEDIEVAL BRITAIN John Gillingham and Ralph A. Griffiths
MODERN ART David Cottington
MODERN IRELAND Senia Pašeta
MOLECULES Philip Ball
MUSIC Nicholas Cook
MYTH Robert A. Segal
NATIONALISM Steven Grosby
NEWTON Robert Iliffe
NIETZSCHE Michael Tanner
NINETEENTH-CENTURY BRITAIN Christopher Harvie and H. C. G. Matthew
NORTHERN IRELAND Marc Mulholland
PARTICLE PHYSICS Frank Close
PAUL E. P. Sanders
PHILOSOPHY Edward Craig

PHILOSOPHY OF LAW Raymond Wacks
PHILOSOPHY OF SCIENCE Samir Okasha
PHOTOGRAPHY Steve Edwards
PLATO Julia Annas
POLITICS Kenneth Minogue
POLITICAL PHILOSOPHY David Miller
POSTCOLONIALISM Robert Young
POSTMODERNISM Christopher Butler
POSTSTRUCTURALISM Catherine Belsey
PREHISTORY Chris Gosden
PRESOCRATIC PHILOSOPHY Catherine Osborne
PSYCHOLOGY Gillian Butler and Freda McManus
PSYCHIATRY Tom Burns
QUANTUM THEORY John Polkinghorne
THE RENAISSANCE Jerry Brotton
RENAISSANCE ART Geraldine A. Johnson
ROMAN BRITAIN Peter Salway
THE ROMAN EMPIRE Christopher Kelly
ROUSSEAU Robert Wokler
RUSSELL A. C. Grayling
RUSSIAN LITERATURE Catriona Kelly
THE RUSSIAN REVOLUTION S. A. Smith
SCHIZOPHRENIA Chris Frith and Eve Johnstone
SCHOPENHAUER Christopher Janaway
SHAKESPEARE Germaine Greer
SIKHISM Eleanor Nesbitt
SOCIAL AND CULTURAL ANTHROPOLOGY John Monaghan and Peter Just
SOCIALISM Michael Newman
SOCIOLOGY Steve Bruce
SOCRATES C. C. W. Taylor
THE SPANISH CIVIL WAR Helen Graham
SPINOZA Roger Scruton
STUART BRITAIN John Morrill
TERRORISM Charles Townshend
THEOLOGY David F. Ford
THE HISTORY OF TIME Leofranc Holford-Strevens
TRAGEDY Adrian Poole
THE TUDORS John Guy
TWENTIETH-CENTURY BRITAIN Kenneth O. Morgan
THE VIKINGS Julian D. Richards
WITTGENSTEIN A. C. Grayling
WORLD MUSIC Philip Bohlman
THE WORLD TRADE ORGANIZATION Amrita Narlikar

Available soon:

AFRICAN HISTORY John Parker and Richard Rathbone
CHAOS Leonard Smith
CHILD DEVELOPMENT Richard Griffin
CITIZENSHIP Richard Bellamy
HIV/AIDS Alan Whiteside
HUMAN RIGHTS Andrew Chapham
RACISM Ali Rattansi

For more information visit our web site

www.oup.co.uk/general/vsi/

Rob Iliffe

NEWTON

A Very Short Introduction

OXFORD
UNIVERSITY PRESS

OXFORD
UNIVERSITY PRESS

Great Clarendon Street, Oxford OX2 6DP

Oxford University Press is a department of the University of Oxford.
It furthers the University's objective of excellence in research, scholarship,
and education by publishing worldwide in

Oxford New York

Auckland Cape Town Dar es Salaam Hong Kong Karachi
Kuala Lumpur Madrid Melbourne Mexico City Nairobi
New Delhi Shanghai Taipei Toronto

With offices in

Argentina Austria Brazil Chile Czech Republic France Greece
Guatemala Hungary Italy Japan Poland Portugal Singapore
South Korea Switzerland Thailand Turkey Ukraine Vietnam

Published in the United States
by Oxford University Press Inc., New York

First published as a Very Short Introduction 2007

British Library Cataloguing in Publication Data

Data available

Library of Congress Cataloging in Publication Data

Data available

Typeset by RefineCatch Ltd, Bungay, Suffolk
Printed in Great Britain by
Ashford Colour Press, Gosport, Hampshire

ISBN 978-0-19-929803-7

1 3 5 7 9 10 8 6 4 2

Acknowledgements

I would like to thank Martin Beagles, John Young, Luciana O'Flaherty, Larry Stewart, and Sarah Dry for commenting on earlier versions of this work, and also for suggesting improvements.

Preface

In Victorian Britain, every schoolboy knew that Sir Isaac Newton was an unrivalled mathematical and scientific genius, and most would have been able to give a basic account of his central discoveries. In optics, Newton found that white light was not a fundamental element within nature but was composed of more basic, primary rays being mixed together. Bodies appeared a particular colour because they had a disposition to reflect or absorb certain colours rather than others. In the realm of mathematics, Newton discovered the binomial theorem for expanding the sum of two variables raised to any given power, as well as the basic laws of calculus. This treated the rate of change of any variable (the shape of a curve or the velocity of a moving object) at any moment, and also offered techniques for measuring areas and volumes under curves (amongst other things). Both his mathematical and optical work took many decades to be fully accepted by contemporaries, the first because his work was shown only to a handful of contemporaries, and the second because many found it hard to reproduce and too revolutionary to be easily grasped.

The crowning glory of Newton's system was contained in his *Principia Mathematica* of 1687, in which he introduced the three laws of motion and the incredible notion of Universal Gravitation – the idea that all massive bodies continuously attracted all other bodies according to a mathematical law. Using completely novel concepts such as 'mass' and 'attraction', Newton announced in his laws of motion (1) that all bodies continued in their state of motion or rest unless affected by some external force; (2) that the change in state of all bodies was proportional to the force that caused that change and took place in the direction exerted by that force; and (3) that to every action there was an equal and opposite reaction. Investigating the consequences of his work in this area formed the basis of celestial mechanics in the 18th century and made possible a new and what we take to be correct physics (special and general relativistic effects excepted) of the Earth and heavens. Not for nothing was Newton held by the vast majority of educated people as the Founder of Reason.

Apart from this, the elites of Victorian Britain grappled with more difficult aspects of Newton's life and work, for it was also known that Sir Isaac was both a committed alchemist and a radical heretic. Incontrovertible evidence also showed that he had behaved in a reprehensible manner towards a number of his contemporaries. Since then, explaining his personality and addressing the problem of reconciling the 'rational' and 'irrational' aspects of his work have continued to challenge historians. Moreover, the fact that many important papers only became available for serious investigation in the 1970s means that a well-balanced picture of his work has only become possible in the last few decades.

Although it has long been known that he had these apparently outlandish interests – which he undoubtedly understood to be more significant than his more 'respectable' pursuits – recent popular biographies of Newton have continually played up these less orthodox elements as if they are being described for the first time. Nevertheless, these books have neither offered new insights, nor do they make use of the astonishing materials that have been made available online in the last few years. Most of these works also make overblown claims about the links between various spheres of Newton's intellectual activity. This introduction aims to redress these problems by taking into account recent scholarly work as well as the newly accessible online transcriptions of writings; as it happens, the Newton that emerges is much stranger than has been visible in recent accounts.

Contents

List of illustrations

The publisher and the author apologise for any errors or omissions in the above list. If contacted it will be pleased to rectify these at the earliest opportunity.

Chapter 1
A national man

Unconscious since late on the previous Saturday evening, Sir Isaac Newton died soon after 1 a.m. on Monday 20 March 1727 at the age of 84. He was attended at his passing by his physician Richard Mead, who later told the great French *philosophe* Voltaire that on his deathbed Newton had confessed he was a virgin. Newton was also looked after in his final hours by his half-niece Catherine and her husband John Conduitt, who had acted as a sort of personal assistant to Newton in his final years. Despite many demands on his time, Conduitt almost single-handedly organized the commemoration of the great man he had come to know, and he heroically managed to supervise the collection of virtually all the significant information that we have concerning Newton's private life. He was responsible for arranging Newton's funeral at Westminster Abbey at the end of March 1727, and he commissioned Alexander Pope to compose the epitaph on Newton's tomb. In the following years he authorized the execution of numerous paintings and busts of his hero by the greatest British and foreign artists of the day.

Over a number of years Conduitt tried to write the definitive 'Life' of Newton, although he never completed the task. He had recorded details of some conversations he had had with Newton but for more detail on Newton's scientific work he asked a number of people to send in their reminiscences. A week after Newton's death he wrote

1. Conduitt's own bust of Newton, executed by J. M. Rysbrack

to Bernard de Fontenelle, Permanent Secretary of the Paris Académie Royale des Sciences, offering to supply the Frenchman with material that he could use in his 'Eloge' of Newton. Conduitt saw this as a chance to secure his relative's reputation in the country that had been most unwilling to recognize Newton's pre-eminence in science and mathematics. It would not be until the late 1730s that Newton's reputation was secure in France, and in the immediate aftermath of his death Conduitt was keen that French and other non-British scholars should be aware of Newton's priority in devising the calculus, an accolade most French scholars still accorded to the German polymath Gottfried Leibniz. Over the summer of 1727, Conduitt worked on a 'Memoir' of Newton, which he sent off to Fontenelle in July.

Conduitt's 'Memoir' gave a factual if adulatory history of Newton's intellectual and moral life, and the latter was described as 'pure & unspotted in thought word & deed'. He was astonishingly humble, exhibited great charitableness and such a sweetness and meekness that he would often shed tears at a sad story. He loved liberty and the Hanoverian regime of George I, 'abhorred and detested' persecution, and mercy to beast and Man was 'the darling topick he loved to dwell upon'. Conduitt included an account of Newton's early development at Cambridge, and added a one-sided version of the priority dispute with Leibniz. Not only had Leibniz not been the first to invent it but he 'never understood it enough to apply it to the system of the Universe which was the great & glorious use Sir Isaac made of it'.

Fontenelle's 'Eloge' was read to the Académie in November 1727. He gave a good account of Newton's scientific and mathematical development, accepting that virtually all of his great discoveries had been made in his early twenties. He disagreed with many of the tenets found in the *Principia*, especially that of the notion of 'attraction', but he was effusive about its overall significance. Although he realized that Newton disagreed with many of the theories of the great French mathematician and philosopher René Descartes, Fontenelle noted that they had both attempted to base science on mathematical foundations, and that both were geniuses in their own time and manner. The Eloge was immediately translated into English, becoming the dominant source for all English-language biographies for over a century.

Other works appeared very quickly, one of which, William Whiston's *Collection of Authentick Records*, was the first text to publicly challenge the view of Newton as a shining white knight. Whiston was Newton's successor as Lucasian Professor at Cambridge but had been ejected from Cambridge in 1710 for espousing heretical religious views similar to those held by Newton. Revealing Newton's radical theological views for the first time, Whiston contrasted Newton's 'cautious Temper and Conduct' with his own 'openness', but remarked that Newton could not hide his own momentous discoveries in theology, 'notwithstanding his prodigiously fearful, cautious, and suspicious Temper'.

Even before he read Whiston, Conduitt was peeved both at the even-handed way with which Fontenelle had compared Newton with Descartes and at his treatment of the priority dispute. He immediately wrote again to a number of pro-Newtonians, pleading in February 1728 that 'As Sir I. Newton was a national man I think every one ought to contribute to a work intended to do him justice.' Of those letters he received in response, the most interesting were two from Humphrey Newton (no relation), who as Newton's amanuensis (secretary) had a unique insight into Newton's behaviour during the years in which he had composed the *Principia* (1684–7). According to Humphrey, Newton would sometimes take 'a sudden stand, turn'd himself about, run up the Stairs, like another Archimedes, with an *eureka*, fall to write on his Desk standing, without giving himself the Leasure to draw a Chair to sit down in'. Newton at this time apparently received only a select band of scholars to his chambers, including John Francis Vigani, a chemistry lecturer at Trinity. Vigani got on well with Newton until, according to Catherine Conduitt, Vigari 'told a loose story about a Nun'.

John Conduitt had already received crucial information from the antiquarian William Stukeley, who had moved to Grantham shortly before Newton's death. Since this was where Newton had attended the local grammar school while lodging with the local apothecary, it was an ideal place to collect information relating to Newton's youth. By 1800 some of the Stukeley material but little from the Conduitt papers had been published. In the early 19th century, however, new information profoundly altered the way people thought of Newton. In 1829 a translation of a recent biography of Newton by Jean-Baptiste Biot revealed that he had suffered a breakdown in the early 1690s. Still more damagingly, in the 1830s a barrage of upsetting evidence emerged from the papers of the first Astronomer Royal, John Flamsteed, which presented a tarnished view of Newton's demeanour. Thereafter, Victorians vied to offer accounts of Newton's life and works. Most importantly, David Brewster's *Memoirs of the Life, Writings and Discoveries of Sir Isaac Newton* (1855), a greatly revised version of his *Life of Sir Isaac Newton* (1831), became the dominant biography for over a century. He tried valiantly to deal with Newton's commitment to alchemy, his unorthodox religious opinions, and his often graceless treatment of both friend and foe, but was ultimately unwilling to recognize the full extent to which Newton fell short of perfection.

In the early 1870s the fifth Lord Portsmouth, a distant descendant of Catherine Conduitt and owner of Newton's papers, generously decided to donate Newton's 'scientific' manuscripts to the nation. A committee was set up at Cambridge University to assess the significance of the collection, and its results were reported in a catalogue of the papers in 1888. The non-scientific papers, including Newton's alchemical and theological writings, were generally deemed of little interest and they remained in the Portsmouth family until they were sold off at Sotheby's in 1936 for the ridiculously small sum of just over £9,000. A syndicate gradually acquired most of the theological papers from dealers, and ultimately they were bought up by the collector Abraham Yahuda, an expert in semitic philology. Yahuda died in 1951 and, although he was an anti-Zionist, his astonishing collection of Newton's papers came into the possession of the Jewish National and University Library in the Hebrew University of Jerusalem after a court case lasting nearly a decade.

The great economist John Maynard Keynes had attended part of the Sotheby sale, and he set his energies towards acquiring all of

Newton's alchemical papers, as well as all the 'personal' papers in the hand of John Conduitt. By 1942, the tercentenary of Newton's birth, Keynes was in possession of the vast majority of Newton's alchemical papers, along with some theological tracts. Although he was preoccupied by the demands of the Second World War, Keynes gave a talk based on these materials as part of the muted tercentenary celebrations. His Newton was far more extraordinary than the person presented by previous biographers, being a 'Judaic monotheist of the School of Maimonides', neither a 'rationalist' nor 'the first and greatest of the modern age of scientists', but

> the last of the magicians, the last of the Babylonians and Sumerians, the last great mind which looked out on the visible and intellectual world with the same eyes as those who began to build our intellectual inheritance rather less than 10,000 years ago.

Newton saw the twin worlds of nature and obscure texts as one giant riddle that could be unravelled by decoding 'certain mystic clues which God had lain about the world to allow a sort of philosopher's treasure hunt to the esoteric brotherhood'. His writings on alchemical and theological topics were, Keynes argued, 'marked by careful learning, accurate method, and extreme sobriety of statement' and were 'just as *sane* as the *Principia*'.

The two most influential scholarly biographies of the late 20th century both made extensive use of manuscript materials. Frank Manuel's *A Portrait of Isaac Newton* of 1968 offers a psychoanalytical account of Newton's personality that is heavily reliant upon the assumption that Newton's unconscious behaviour expressed itself 'primarily in situations of love and hate'. According to Manuel, the source of Newton's psychic problems lay in the fact that she remarried when Newton was only 3 years old. Having already lost his biological father, who died only months before he was born, Newton became hostile to his stepfather and devoted himself to the one Father he could really recognize – God. Manuel showed how the traumatic experiences of Newton's youth were internalized, and the brilliant but tormented young Puritan became the ageing despot of the early 18th century.

In his more orthodox *Never at Rest: A Scientific Biography of Isaac Newton* of 1980, Richard S. Westfall took Newton's work as the central aspect of his life. Drawing from the full range of Newton's manuscripts that were now available to scholars, his 'scientific biography' engaged with every aspect of Newton's intellectual interests, although his scientific career 'furnishes the central theme'. While he deals ably with Newton's intellectual accomplishments, it is apparent that Westfall's great admiration for this part of Newton's life does not extend to his personal conduct.

Ultimately Westfall came to loathe the man whose works he had studied for over 2 decades. He was not the first to feel this way about the Great Man.

Chapter 2
Playing philosophically

According to the calendar then in use in England, Newton was born on Christmas Day 1642 (4 January 1643 in most of Continental Europe). The first decade of his life witnessed the horror of the civil wars between parliamentary and royalist forces in the 1640s, culminating in the beheading of Charles I in January 1649. His uncle and stepfather were rectors of local parishes, and they seem to have existed without much harassment from the church authorities convened by Parliament to check for religious 'abuses'. In his second decade he lived under the radical Protestant Commonwealth, which was replaced in 1660 when Charles II was restored to the throne. Newton was born into a relatively prosperous family and was brought up in a devout atmosphere. His father, also Isaac, was a yeoman farmer who in December 1639 inherited both land and a handsome manor in the Lincolnshire parish of Woolsthorpe. His mother, Hannah Ayscough, came from the lower gentry and (as was common for the period) seems to have been educated at only a rudimentary level. Nevertheless, her brother William had graduated from Trinity College Cambridge in the 1630s and would be influential in directing Newton to the same institution.

Newton's father, apparently unable to sign his name, died in early October 1642, almost three months before the birth of his son. Newton told Conduitt that he had been a tiny and sick baby, thought to be unlikely to survive; two women sent to get help from a local gentlewoman stopped to sit down on the way there, as they were certain the baby would be dead on their return. Surviving against the odds, Newton was brought up by his mother until the age of 3, when she was approached with an offer of marriage by Barnabas Smith, an ageing vicar of a local parish. Smith was wealthy, and they married in January 1646 after he had promised to leave some land to her first born. Spending most of her time with her new spouse, she produced three more children before his death in 1653 (one of whom would be the mother of Catherine Conduitt). Although John Conduitt waxed lyrical about Hannah's general virtues, and was careful to point out that she was 'an indulgent parent' to all the children, he emphasized that young Isaac was her favourite. Whatever the truth of this, Newton's own evidence indicates that, as a teenager, he had an extremely difficult relationship with his mother, and historians have always found it difficult to make Conduitt's account tally with the fact that for seven years Newton was effectively left in Woolsthorpe to be brought up by his maternal grandmother.

Newton went to two local schools until he was 12, after which he went to Grantham Grammar School. Here he lodged with a local apothecary, Joseph Clark, whose shop proved to be a great source of information. A descendant of Clark told William Stukeley that Newton showed an immense interest in the abundant medicines and chemicals, and Stukeley noted that he spent a great deal of time gathering herbs, probably learning about their properties from Clark's apprentices. Newton lived with Clark's stepchildren, one of whom, Catherine, who grew up to be a Mrs Vincent, provided abundant information about the prodigy. Everyone Stukeley met recounted 'the extraordinary pregnancy of his genius' for building machines and told him 'that instead of playing among the other boys, when from school, he always busyed himself at home, in making knickknacks of divers sorts, & models in wood, of whatever his fancy led him to'. Mrs Vincent, allegedly the object of amorous attention from the young inventor, recorded that his schoolfellows were 'not very affectionate' towards him, aware 'that he had more ingenuity' than they did. Instead, little Isaac was 'always, a sober, silent, thinking lad', who never played with boys but who would occasionally make dolls house furniture for the girls 'to set their babys, and trinkets on'.

Newton built up 'a whole shop of tools' in Grantham, spending all the money his mother gave him on saws, chisels, hatchets, hammers and the like, 'which he would use with as much dexterity, as if he had been brought up to the trade'. Many of the machines described by Mrs Vincent and others had been originally set out in a book by John Bate entitled *Mysteries of Nature and Art*, part of an extremely popular genre of 'mathematical magic' books that contained numerous recipes and drawings of machines. Newton was already unwilling simply to appropriate information without developing it in a dramatic fashion. Not content with reproducing a simple windmill described in Bate, he went to see a real version being constructed in a neighbouring village, 'was daily with the workmen' and 'obtain'd so exact a notion of the mechanism of it, that he made a true, & perfect model of it'. He went beyond his prototype and adjusted the mechanism so that the sails were powered by a mouse, which drove a wheel in its efforts to reach some corn. While Stukeley's informants disagreed as to its exact mechanism, they concurred that people would come from miles around to see Isaac's 'mouse miller'. Stukeley perceptively noted that 'ludicrous' (i.e. playful) devices commonly grabbed his attention. Apart from the mouse miller and the dolls' furniture, Newton examined the fabric and dimensions of a simple kite, built a better example, and attached a candle-lit lantern to it, frightening the countryfolk and giving them much to discuss as they drank their beer.

As in the cases of the windmill and the kite, Newton made a wooden clock and then immediately built a better one. This improved version, which had a dial, was powered by a steady trickle of water that he supplied each morning, and was made from a box given to him by Humphrey Babington. Babington, the brother of Mrs Clark (a close friend of Hannah Smith), had been ejected from Trinity

College for refusing to take the engagement oath of allegiance to the Commonwealth, and would play a significant role in Newton's life over the following decades. Extending his virtuosity still further, Newton graduated to complex sundials, turning various features of Clark's house into different sorts of clock and, according to Stukeley, 'showing the greatness, & extent of his thought by drawing long lines, tying long strings with running balls upon them; driving pegs into the walls, to mark hours, half hours & quarters'. He made an 'almanac' of these lines, 'knowing the day of the month by them; the suns entry into signs, the equinoxes, & solstices'. 'Isaac's dials', like many of his other accomplishments, became well known in the parish. Perhaps the greatest of his juvenile achievements, Stukeley believed that these were the origins of his fascination for heavenly motions.

Newton also excelled in artistic pursuits, such as drawing and even the composing of poetry, though his penchant for verse would prove temporary. He covered the walls of his attic room with charcoal drawings of animals, men, plants and mathematical figures, and scratched his name into the shelves. In the middle of the 20th century, geometrical drawings, undoubtedly by Newton, were discovered etched onto the stonework of Woolsthorpe Manor.

Newton's artistic bent at this time can be gauged by a series of notes on Bate's book, entered into a notebook that he purchased in 1659. These notes attest to Newton's concern with the practical aspects of drawing, and also his interest in producing a wide variety of coloured inks and paints, whether from animals, vegetables, and minerals, or by mixing pre-existing colours. Just over a decade later, the last of these topics would make him famous. Other instructions concerned how to make fishbait and different ways, not all of them overly complicated, of catching birds by making them drunk. Bate's book also contained recipes for universal salves and ointments, a number of which Newton noted down. Indeed, one of the few things later recalled by John Wickins, his roommate of 20 years at Cambridge, was that Newton would often take a grisly self-prepared

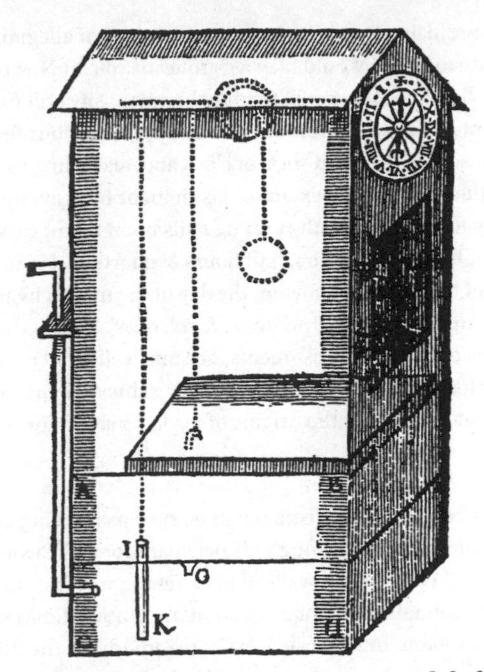

2. The source for Newton's design for a water-powered clock, from John Bate's *Mysteries of Nature and Art*

concoction ('Lucatello's balsam') as a preservative. Some notes came from John Wilkins's *Mathematical Magick*, a popular work that purveyed similar information to Bate, while other entries in the notebook concerned different ways to produce perpetual motion, a topic of extreme interest in the following decades.

This immersion in worlds of practical ingenuity not only offered portents of his great future, but led directly to it. Indeed, Stukeley gave a superb account of how Newton's early obsessions related to his later triumphs. He pointed out that Newton's early mastery at using mechanical tools, along with his expertise in drawing and designing, was extremely useful for his experimental skill and 'prepar'd for him a solid foundation to exercise his strong reasoning facultys upon'. Uniquely Newton had all the qualities for becoming a great natural philosopher, such as 'profound judgement', 'invincible constancy, & perseverance in finding out his solutions', 'a vast strength of mind, in protracting his reasonings [and] his chain of deductions', and an 'incomparable skill in algebraic, & the like methods of notation'. Like all children he was an imitator, but for Stukeley 'he was in reality born a philosopher. Learning, & accident, & industry pointed out to his discerning eye some few, simple & universal truths', which he gradually extended 'till he unfolded the œconomy of the macrocosm'.

A godly child

Absorbed as he was in making his devices, the gifted country boy was a deeply unhappy youth. Late in May 1662 he recorded a list in shorthand of all the sins he had committed in the previous decade, and for a short time he noted down all the misdemeanours committed while at Cambridge. The term 'Puritan' is strictly false as a description of Newton's religious doctrine but the radical Protestant ethical values associated with this term accurately describe the person who appears in the entries. Many of the sins cover activities performed on the Sabbath ('Thy day'), when godly Christians were supposed to rest. On various Sundays in the 1650s, Newton read a frivolous book, ate an apple in chapel, and made a feather, a clock, a mousetrap, some rope, and in the evening some pies. He confessed to 'idle discourse' on God's day, so that it is not surprising that he also carelessly heard and committed to memory various sermons, while he also recorded that he completely missed chapel on one occasion. Sometimes he had set his heart on learning and money more than on God, preferring 'worldly things' instead, and indeed many of the sins recall his failure to live as a godly man. 'Not living according to my belief' and 'neglecting to pray', he had become distant from God, failing to love God for Himself and failing to 'long' for God's ordinances.

Some episodes were those common to any teenager in his village.

He put a pin in another boy's hat to 'prick' him, refused to come home when his mother told him to, and lied to his mother and grandmother about having a crossbow. At other times, he 'fell out' with servants. Food crimes were also prominent: he stole cherry cobs from Edward Storer, Clark's stepson, and pilfered plums and sugar from his mother's foodbox. He even confessed to gluttony while he was ill, and indeed the first entries in the short list of sins committed when he was a student at Cambridge were for the same offence. Other comments in the first list portray darker elements of his psyche. He punched one of his sisters, struck 'many', and beat up Arthur Storer, Edward's brother. The precise meaning of 'Having unclean thoughts words and actions and dreams' in Newton's list is unclear, as is his lament that he had used 'unlawful means' to bring himself out of 'distress'. Real loathing shows through his recollection of 'wishing death and hoping it to some', and most horrifying of all is the distant memory of having threatened to burn his stepfather and mother along with their house. Newton also compiled a list of common words arranged alphabetically in Francis Gregory's *Nomenclatura brevis reformata* of 1651. To terms like 'Father', 'Wife', and 'Widdow', Newton added words such as 'Fornicator' and 'Whoore' not found in Gregory, expressions that perhaps refer to his view of his mother and stepfather.

Newton's anger manifested itself in other areas of his life. According to Conduitt, who knew him well, resentment and the desire to emulate had been the forces propelling Newton to outdo all others at the start of his academic career. Newton often told him a story about his early days at the grammar school when he was at the bottom of the class, a narrative that is possibly connected with his 'confession' about beating Arthur Storer. One day he was kicked in the stomach on his way to school. After lessons had ended he fought in the churchyard with his assailant, and although Newton 'was not so lusty as his antagonist he had so much more spirit & resolution that he beat him till he declared he would fight no more'. Later, the schoolmaster's son goaded him into forcing his antagonist's face into the side of the church. After this, Newton strove to outdo his opponent in learning, not stopping until he had risen above him in the pecking order. Inexorably, he rose to become top of the school.

His extracurricular activities had an adverse effect on his schooling but such was his ability that he could resume his academic work and outperform his schoolfellows whenever he wanted. Stukeley noted that 'dull boys were sometimes put over him, in form, but this always excited him to redouble his pains, to overtake them'. The headmaster of the school, John Stokes, seems to have spotted Newton's talent at an early stage, but could not coax the lad away from his hammers and chisels. However, in the latter half of 1659 his mother decided to pull him out of school to run the family estate. Despite being put in the care of a trusty servant, his obsession with building waterwheels and other models and a capacity to be lost in his books made Newton completely unsuitable for the task. The sheep and cows he was supposed to be looking after strayed into neighbouring fields, and records show that he was fined for this in October of the same year. He could barely remember to eat and, according to Stukeley, 'philosophy absorbed all his thoughts'.

It is at this point that narratives of Newton's development begin to portray him as an unworldly scholar rather than as a gifted mechanic. Later, a number of different pieces of evidence indicate that he became famous for his unworldly or 'insensate' behaviour when he went to Cambridge. A hopeless manager of his family's affairs, he would bribe the servant to act on his behalf, and he would find scholarly refuge in the attic where he had lodged while at the school, engrossed in a pile of medical and scientific tomes that had been left there. On other occasions, he would simply lie under a hedge or a tree and read a book. Once Newton's horse slipped his bridle, and he walked on unawares for miles, engrossed in a book he was reading. His mother was 'not a little offended at his bookishness', while the servants called him 'a silly boy' who 'would never be good for any thing'.

To the rescue came Stokes, who told Hannah that Newton's immense talent should not be buried in 'rustic business'. He saw 'the uncommon capacity of the lad, & admired his surprising inventions, the dexterity of his hand, as well as his wonderful penetration, far beyond his years', telling his mother that he 'would become a very extraordinary man'. Stokes offered to let him board for free, possibly a key factor in Hannah allowing her son to go back to the grammar school to prepare for university. Returning there in the autumn of 1660, he received extra tuition in Latin and Greek, and on his final day was given a rousing send-off by Stokes, allegedly driving the rest of the school to tears. Stukeley noted that no such sentiment was felt by the servants, who declared him 'fit for nothing but the Versity'.

Trinity

By this time it had already been decided that he would go to Trinity College Cambridge, the most prestigious college in England. The combined forces of William Ayscough and Humphrey Babington, newly restored as a fellow, were probably decisive in sending Newton there. Newton arrived in Cambridge on 5 June 1661 in the relatively menial position of 'subsizar', a lowly status strangely out of keeping with the wealth that his mother commanded. Subsizars, who had to pay for their own food and also to attend lectures, were effectively servants of fellows or wealthy students, and it is possible that Newton worked in this position, however notionally, for Babington. Both town and gown had reacted quickly and positively to the restoration of Charles II the previous spring, and in the most senior positions royalist sympathizers had replaced Commonwealth appointees. The Anglican scholar John Pearson, author of the highly influential *Exposition of the Creed* in 1659, became master in 1662, and under him the college emphasized more traditional forms of scholarship and in particular theological study.

Evidence from a small notebook sheds some light on how Newton spent his time and money as an undergraduate. Early entries show his purchase of basic equipment such as books, paper, pen and ink, and the ordinary materials for living in 17th-century student accommodation, such as clothes, shoes, candles, a lock for his desk, a carpet for his room, and a chamberpot. He bought a watch, a chessboard and later a set of chess pieces (according to Catherine Conduitt, he became extremely proficient at board games), and paid seven pence as his yearly subscription for access to the tennis court. The entry 'to balls & barges', repeated later on, indicates that not every moment was spent in study in his first year there. Indeed, he created a separate list of 'frivolous' and 'wasteful' expenses, including the purchase of cherries, beer, marmalade, custard tarts, cake, milk, butter, and cheese. Later, he graduated to apples, pears, and stewed prunes.

Very quickly – and uniquely among undergraduates for whom records survive – Newton began to lend money to his bedmaker and to fellow students, many of them 'pensioners' who occupied a social rank in the college somewhat higher than his. Most recipients of Newton's generosity paid him back, as indicated by a cross through the relevant record. At some point, probably in 1663, Newton met another pensioner, John Wickins (whose son Nicholas recorded that his father had found Newton 'solitary and dejected'), and they decided to room together. Wickins would occasionally act as an amanuensis for Newton until he left Cambridge in 1683 to take up a position in the church. Nick Wickins was told by his father that Newton would forget his food when working and in the morning would arise 'in a pleasant manner with the satisfaction of having found out some Proposition; without any concern for, or seeming want of his Nights sleep'. If Newton's recollections are correct, in the same year he met Wickins, he became fascinated by judicial astrology – the assessment of an individual's future prospects on the basis of studying the positions of the stars and planets – and bought a book on the topic. It was as a result of being dissatisfied with this that he turned the following year to the mathematics of Euclid, only to reject it as trivially obvious.

He probably attended the initial Lucasian mathematical lectures of Isaac Barrow, the first holder of the chair, in March 1664 – and the professor may have noted a particularly attentive student in the audience. In the month after Barrow's inaugural lecture, Trinity held one of its periodic scholarship competitions, which Newton entered. As he told the story later, Barrow was his examiner and – never imagining that the young student had ventured into Descartes's formidable *Géometrie*, a feat that Newton was apparently too modest to admit – was dismayed by Newton's lack of knowledge of Euclid. Newton got the scholarship nevertheless, and thus became entitled to a number of privileges. Early the following year, at about the same time as he discovered the generalized binomial theorem, he was forced to undertake a protracted examination in more standard learning to qualify for his Bachelor of Arts degree. A later tradition held that he almost failed this exam, although the story may be a confusion of this event with the scholarship examination of the previous year.

Plague devastated various parts of England in the middle of 1665 and, along with most other students, Newton returned home some time in late July or early August. Having come back to Cambridge in March 1666, he continued to lend money to many of the same students as before, but when a resurgence of the plague occurred in early summer, he again sought refuge in Lincolnshire. Much of his most innovative work was produced here, probably at the home of Babington in Boothby Pagnell. On 20 March 1667 he received £10 from his mother, who gave him the same amount when he returned to Cambridge in the following month. Over the next year, he spent much of this money, as well as funds repaid by debtors, on equipment for grinding tools and performing experiments, three pairs of shoes, losing at cards (twice), drinking at a tavern (twice), some early volumes of the *Philosophical Transactions*, Thomas Sprat's recently published *History of the Royal Society*, and some oranges for his sister. In September he entered another competition, this time for a college fellowship. Whether because of support from Babington or Barrow, or simply because his brilliance and dedication to scholarship shone through during the four days of the oral examination, Newton was elected minor fellow.

Evidently, this also implied that he was expert in the sort of theological scholarship demanded by Pearson and, as a consequence of his election, he swore to make theology the focus of his studies and to take holy orders – or resign. Soon afterwards he moved to a new room, and revamped it to suit his tastes. In July 1668 he was made a Master of Arts, allowing him to progress to the position of major fellow of the college. He spent more money on material for his gown, and purchased an expensive hat, a suit, some leather carpets, a couch (jointly bought with Wickins), and some materials for a new featherbed. He also bought three prisms at one shilling each, along with 'glasses', presumably for chemical experiments, while in late summer he made his first trip to London. His reputation would soon follow him.

Chapter 3
The marvellous years

The first decades of the 17th century witnessed an exponential growth in the understanding of the Earth and heavens, a process usually referred to as the Scientific Revolution. The older reliance on the philosophy of Aristotle was fast waning in universities, although across Europe Aristotelian natural philosophy and ethics would be routinely taught at undergraduate level until the end of the century. In the Aristotelian system of natural philosophy, the movements of bodies were explained 'causally' in terms of the amount of the four elements (earth, water, air, fire) that they possessed, and objects moved up or down to their 'natural' place depending on the preponderance of given elements of which they were composed. Natural philosophy was routinely contrasted with mathematics or 'mixed mathematical' subjects such as optics, hydrostatics, and harmonics, where numbers could be applied to measurable external quantities such as length or duration. All this took place in a cosmos where the Earth was planted at the centre, surrounded by the Sun and the planets.

The first dramatic change took place in astronomy, where despite official opposition from the Catholic Church and from many Protestant denominations, the Copernican heliocentric (sun-centred) system gained new converts. Between 1596 and 1610, there was an astronomical revolution galvanized by the work of Johannes Kepler and Galileo Galilei. Kepler's *Mysterium*

Cosmgraphicum of 1596 posited a heliocentric system in which the distances between the planets could be determined by inscribing the orbits of the planets inside regular solids. He published a magnetic theory of planetary motion in his great *Astronomia Nova* of 1609, a treatise that contained the first two of what were later known as Kepler's Laws (that planets move in ellipses, and that with respect to the Sun, located at one of the foci of a particular orbit, all planets swept out equal areas in equal times).

In 1609 Galileo developed a combination of lenses into a device that allowed him to magnify objects. He turned this 'telescope' to the heavens and realized that Jupiter had a series of satellites that orbited it, just as the planets orbited the Sun. In his short *Sidereus Nuncius* of 1610, he also announced that the Moon had mountains and valleys, and that the Milky Way was composed of thousands of stars. In 1613 he would further challenge the standard view, which held that the heavens were 'incorruptible', by demonstrating that the Sun had spots. Kepler would add his Third Law in his *Harmonice Mundi* in 1619, which stated that for any planetary orbit, the ratio between the cube of the mean radius of the planet from the Sun, and the square of its period of revolution, was constant. While Galileo's discoveries effectively demolished belief in the perfection of the heavens, Kepler's laws would be of central importance for Newton in demonstrating key propositions in the *Principia*.

Galileo's contribution to 17th-century science did not end with his work in astronomy. In 1632 he bravely published his *Dialogo sopra i due Massimi Sistemi*, a work which attempted to prove the Copernican system of the world. For this he was placed under house arrest until the end of his life in 1642, although his brilliant *Discorsi e Demonstrazioni Matematiche Intorno a Due Nuove Scienze* appeared in 1638. Aristotle had assumed that projected bodies first experienced 'violent' motion, which was then taken over by the 'natural' motion that drove the earthy particles of the object downwards to their natural place. He had also argued that bodies fell at speeds proportional to their weight. Instead, Galileo announced in the *Discorsi* that the trajectory of projectiles was parabolic, while the vertical component of a body near the surface of the Earth could be expressed as a law according to which – for bodies of any weight, or 'bulk' – the total distance fallen vertically is proportional to the square of the time taken. He also made it clear, again in opposition to the entire Aristotelian project, that the physical causes of gravity were unimportant, and indeed, would be extremely difficult to uncover. In showing that a number of phenomena in the terrestrial sphere were mathematizable, Galileo laid the basis for the modern science of mechanics. Newton's great triumph – expressed in his momentous work of the same name – was to show that 'mathematical principles' were at the basis of many more natural phenomena.

Another essential dimension of modern science was outlined in the work of Francis Bacon. At the same time that Galileo and Kepler were developing astronomy and mechanics, Bacon was promoting the idea that the proper way to understand nature was to directly engage with it rather than approach it through the medium of Aristotelian (or any other) texts. Arguing that a collaborative project was the only way to achieve progress in natural philosophy, Bacon pointed to the recent discoveries of America and the Pacific Ocean and praised the advances made by arts and trades. Observations of disparate facts would increase knowledge of the visible world while well-designed experiments would break the natural world down into its constituent parts and convey information about nature's real secrets. Bacon even praised the way in which alchemists were prepared to analyse nature, though he lamented their closeted lifestyles and opaque jargon.

Not all anti-Aristotelians agreed that Galileo's project was the proper way to uncover scientific truths. René Descartes developed a sophisticated account of the sorts of nano-structures underlying the physical world. He assumed that the machine-like phenomena that existed in the world around us also operated at the invisible level. In his mechanical philosophy, an unseen microworld was populated with hooks and screws, which made elements cohere. Large-scale phenomena such as magnetism, heat, gravity, and electricity were explained through the activity of a giant solar 'vortex', which by spewing out various sorts of matter had major effects on terrestrial phenomena. Descartes shared Galileo's anti-Aristotelianism (and secretly, his and Kepler's Copernicanism) but he accused the Italian of 'building without foundation', arguing that scientific explanations needed to be couched in terms of the micro-mechanical building-blocks of nature. This, as we shall see, was the most influential work for the young Newton, although it was soon the object of his critical animus.

A mathematical tyro

At first, Newton's education, was that of a standard Cambridge undergraduate, and he was required to read a substantial amount of the prescribed theological and Aristotelian literature. It may well have been the Lucasian lectures of Barrow in the spring of 1664 that spurred his interest in serious mathematics, and Newton later recorded that he read William Oughtred's *Clavis Mathematicae* and Descartes's *Géometrie* about the time that Barrow began lecturing. In the winter of 1664–5 he closely studied the analytic mathematics of Descartes (and the commentary in his edition of the latter's *Géometrie* by the Dutch mathematician Frans Van Schooten), François Viète's work on algebra, and John Wallis's 'method of indivisibles'. Using what we call Cartesian co-ordinate geometry, he mastered the equations that defined the various conic sections (circles, parabolas, ellipses, and hyperbolas). Although he had initially underestimated the achievement of Euclid in his *Elements*, he would later revere the classical accomplishments of Euclid and Apollonius, taking their approach to be the template for doing mathematical work.

Towards the end of 1664 Newton found out how to measure the 'crookedness' or slope of a curve at any point. This was known as the

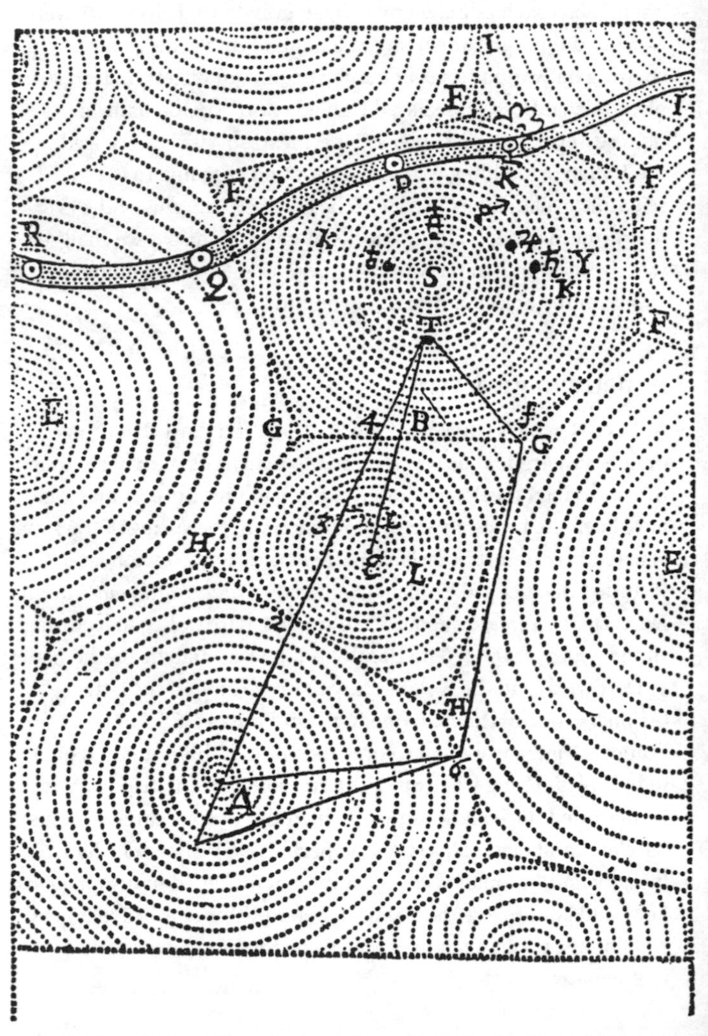

3. Cartesian vortices: the solar system, surrounding the Sun, S, being bounded by FFFFGG. Other systems have stars at their centre.

‘problem of tangents’, and was being developed by mathematicians such as James Gregory and René François de Sluse. Newton soon built on an approach formulated by Descartes, by which the ‘normal’ to a curve (i.e. the line perpendicular to the tangents) could be determined by finding the radius of curvature of a single large circle at the point at which it touches the curve. Newton took the ‘normals’ between two close points, allowing the distance between them to become arbitrarily small. He could now find the tangent to any point on equations that ‘expressed’ any conic section, as well as the maxima and minima of related equations. He generalized the procedure to express the basic elements of what we call differentiation, by which the slope of the tangent represents the rate of change of a curve at any point.

As early as the winter of 1663–4 he had begun to read Wallis's analysis of the ways in which areas under sections of a curve could be found by dividing the space into infinitely small sections. By the time Wallis published his *Arithmetica Infinitorum* in 1655, it was known that for basic equations $x = y^n$, the area under the curve between 0 and a was $a^{n+1}/n + 1$. This was known as ‘squaring’ or ‘quadratures’, and was the embryonic form of what we now call integration. More complex equations demanded different techniques such as the use of infinite series, which allowed an approximation to a final value as a series of terms reached a limit. Wallis had developed this idea, squaring the parabola and hyperbola and discovering a series of terms that approached the value of π.

Newton read Wallis carefully in the winter of 1664–5 and offered alternative techniques for achieving the same results. Soon he refined Wallis's technique so as to consider the quadratures of curves with fractional powers (i.e. involving square, cube, and other roots). He went beyond Wallis by finding the correct series to square the circle and as a result of extending the insights gained from this success, he eventually discovered the generalized binomial theorem (i.e. for fractional as well as integral powers) for expanding any equation of the form $(a + x)^{n/m}$, publicly announced for the first time in a letter to Leibniz in 1676.

Early in 1665 Newton understood generally that the techniques of tangents and of quadratures were inverse operations, that is, he had the fundamental theorem of the calculus. By late 1665, and possibly in imitation of Barrow, he was routinely treating curves as points that carved out lines in a virtual space under certain conditions, and he referred to the 'velocities' that points experienced in given moments of time. This was what he called the 'fluxional' calculus, because the values of points on the curve 'flowed' from one point to the next. Areas under curves could now be treated not just as sums of infinitely small segments, but as areas 'kinematically' created by considering the space traversed by lines connecting a moving point to corresponding values directly beneath the point on the x-axis. Most of this brilliant work was systematized in an extraordinary essay of October 1666, a treatise that marked him out as the leading mathematician in the world.

The apple

The story that Newton was prompted by a falling apple to think of comparing the force that caused the apple to fall with that required to keep the Moon in its orbit is arguably the best known tale in the history of science. Whether or not it is true, at the same time as he made his mathematical discoveries he was branching out into an extraordinary series of researches into mechanics that would make him the first to unite the forces governing motions on earth and in the heavens. By his own admission, Newton began his novel insights by discovering the law by which a revolving body was kept in its orbit. He soon wrote out a series of laws of motion, many of which he would recall (and develop) when he wrote the *Principia* 20 years later. In a notebook entitled the 'Waste Book', in early 1665 he wrote out over a hundred axioms of motion. These embraced the basic notion of inertia while he also invoked a metaphysical justification for holding that the effects of impacts had to be equal to their cause, an embryonic version of what would be the third Law of Motion in his *Principia*. Taking into account the bulk of a body and its velocity, Newton's exquisite analysis led to a law stating the conservation of momentum (mv) before and after impact.

Next, Newton adroitly investigated the path of a body being bounced from the sides of an enclosed square, imagining that the sum total of the four impacts exerted by each side of the square was analogous and equal to the total force that would be required to keep a body in orbit around a central point. On the assumption that the number of sides exerting an impact could be made infinitely large (so that it was a circle), he concluded that the total force required to keep the body moving in a circle in one revolution was 'to the force of the bodies motion as all those sides [i.e. the circumference of the circle] to the radius'. If the 'force of the bodies motion' was mv, then the total force exerted in one revolution was $2\pi mv$. If the time taken for one revolution was $2\pi r/v$, then the force divided by the time, expressing the *force acting on a revolving body at a given instant*, was mv^2/r. This seminal result in the development of mechanics was first published by Christiaan Huygens in 1673, although years before this Newton had already used it to go beyond what Huygens would achieve.

Newton now realized that he could attack a problem first raised by Galileo, namely the ratio between the force that keeps an object on Earth (gravity) with the 'centrifugal force', the tendency of the same body to be flung off into space by the Earth's rotation. For the first he independently derived g, the acceleration due to gravity. For the second he determined that centrifugal force would propel a body in one revolution of the Earth through the length $2\pi^2 r$, and with a value for the size of the Earth he concluded that the force of gravity was about 350 times stronger than centrifugal force (in one second gravity would make a body descend 16 feet, while centrifugal force would make it travel just over half an inch).

Perhaps influenced by seeing the fall of the apple, in the late 1660s Newton compared the tendency of the Moon to leave the Earth with the force of gravity at the Earth's surface, a problem suggested by Galileo. By using a figure for the size of the Earth that made the Moon about 60 Earth radii (i.e. the distance from the centre of the Earth to the equator) distant, he deduced that the tendency of an object to recede from the Earth's equator (its centrifugal force) was about 12 and a half times that of the Moon to recede from the Earth. If the regularity of the Moon's orbit required the centrifugal force to balance the centrally directed attraction exerted by the Earth, then the centrifugal force of the Moon was equal to 350×12.5 (= 4325) times the gravitational pull of the Earth at its surface.

In the same manuscript in which he made this calculation, Newton derived the inverse-square ($1/r^2$) distance law for the force exerted on a revolving body by inserting his own law for the force of a revolving body into Kepler's Third Law. Newton would later recall that his figure for the force keeping the Moon in its orbit (i.e. 4325) 'answered pretty nearly' to that produced by taking into account the square of the distance between the Moon and the Earth ($60^2 = 3600$) demanded by the inverse-square law. At this point he attributed the difference between these results to the effects of a terrestrial vortex; later he would realize that it was due to an incorrect measurement for the size of the Earth. He would also come to see this incredible effort as evidence for his priority in devising Universal Gravitation. However, amazing as it was, it lacked many of the elements of his great theory.

Philosophical questions

These interests by no means exhausted Newton's scientific fertility, and in another notebook he took a series of notes from Aristotelian texts and from commentaries on them. These covered subjects in the general curriculum that would be studied by any student in a European university, such as ethics, logic, rhetoric, and natural philosophy. At some point, probably late in 1664, he stopped taking excerpts from the Aristotelian textbooks and entered a series of notes and philosophical queries under the heading 'Certain Philosophical Questions'. Above the title he noted a common phrase that in English reads 'Plato is my friend, Aristotle is my friend, but truth is a greater friend'.

The initial entries in the 'Philosophical Questions' notebook were composed under headings concerning the nature of matter, the reason why some tiny bodies 'cohered' together to form larger bodies, the nature of heat and cold, and the question of why some bodies fell and some rose. He made compelling criticisms of conventional views, and indeed the general topics on which he commented would be the focus of his interest for the rest of his life. The earliest entries have a metaphysical flavour to them, which is very different from the more experimental approach he would soon adopt. Regarding the nature of matter, for example, he followed Henry More in the latter's *Immortality of the Soul* (1659) and noted that the primary building-blocks of the physical world must be atoms. Unlike 'mathematicall points', matter could not be divided into infinity, since an aggregation of infinitely small parts, no matter how small they are, could not make a finite object. Regarding cohesion, Newton drew on the Cartesian assumption that a solar 'vortex' spewed out a rarefied matter that gave rise to the atmosphere; this in turn 'pressed down' on the Earth causing 'a close crouding of all the matter in the world'.

Newton would remain committed to a Cartesian-style vortex until the early 1680s. The finest parts of the vortex he termed the 'ethereall mater', although later he would use the word 'aether' to distinguish this pervasive but undetectable medium from the coarser 'air'. He queried whether the agitation of the vortex caused objects to heat up, and also wondered whether heat was caused by air moved by light, or directly by light itself. He also posed the question of whether water could be made to freeze by removing its heat inside Boyle's air-pump (which evacuated or compressed air inside a glass chamber). As for the downward motion of the matter that caused gravity, it must rise again in a different form because (*a*) otherwise the underground cavities of the Earth would swell, and (*b*) the upward rising matter would cancel out the downward, and there would be no gravity. He also argued that the ascending matter had to be 'grosser' than the descending matter, otherwise it would impact upon more (i.e. internal) 'parts' of large bodies and hence give a more powerful upward than downward force. This interest in a cyclical cosmos never waned, and its significance can be seen in his later alchemical and scientific work.

Even heavenly phenomena could be investigated by experiment. Notes from Descartes's *Principia* about the nature of comets were followed immediately by Newton's own observations of the comet of December 1664, an event whose demands on his time and energy he would later remember as making him 'disordered'. Newton noted that the comet moved north 'against the streame of the Vortex' and he proposed extraordinary experiments for testing the possible effects of the lunar vortex. Did the Moon's influence cause tides? No, he suggested initially, because they would be least when there was a new moon but this did not happen. Nevertheless, it might be possible to get a tube of mercury or water and see whether the height of liquid in the tube was affected by the various aspects of the Moon.

At each point Newton proposed experiments for deciding central philosophical questions. No other undergraduate did anything like it. He put forward a series of tests for determining the specific gravity of different elements, and also for ascertaining whether the weight of bodies was affected by being heated or cooled, or by being moved to different places or heights. Fascinatingly, given his theory of gravity, he also queried whether the 'rays' of gravity could be reflected or refracted like light. If some of the gravity rays could be made to strike a horizontal wheel with slats angled at a particular degree to make it turn like a windmill, or if they were only allowed to make contact with one half of a vertical wheel in order to make it revolve, then maybe there could be perpetual motion. Similarly, he posed a series of queries elsewhere in the notebook for deriving perpetual work from magnetical rays. Perhaps, by transmitting these rays, a magnet could produce revolutions in a red-hot iron shaped into sails like those of a windmill? Presumably to test these views, he purchased a high-quality magnet in 1667 and a short time later performed a series of highly original experiments with magnetic filings.

Questions about the nature of air and water were again prompted by his reading of Descartes's *Principia Philosophiae*, and the latter's account of the micro-structures of hard and soft bodies took up much of his energy. Here, as elsewhere, Newton proposed the use of Boyle's air-pump to resolve abstruse theoretical conjectures, many of them concerning the aether. The refraction of light, for example, did take place in an evacuated air-pump, so that it had to be caused by 'the same subtile matter in the aire & in vacuo'. But was the extent of refraction the same in different kinds of glass? Boyle had not considered this, but Newton did, and indeed he had access to an air-pump in Christ's College.

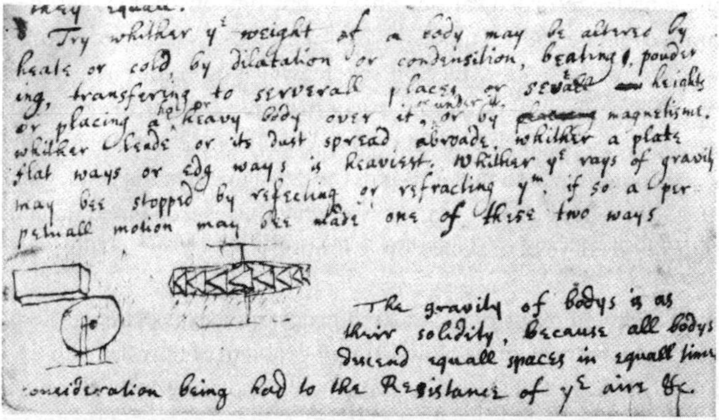

4. Two ideas for perpetual motion machines powered by gravitational waves, from Newton's Trinity College 'Philosophical Questions' notebook

Of mind and body

Many entries in the 'Philosophical Questions' notebook are concerned with the nature and precise location of the soul, and the respective roles that the internal, subjective mind and external bodies played in experience. From the beginning Newton was fascinated by what we would call the mind–body problem, and also by the fact that different people had varied reactions to the same cause. Under the heading 'Of sympathie or antipathie' he noted that

> To one pallate that is sweete which is bitter to another. The same thing smells gratefully to one displeasingly to another . . . Objects of sight move not some but cast others into an extasie. Musicall aires are not heard by all with alike pleasure. The like of touching.

In another section entitled 'Of Sensation' (in notes taken from More's *Immortality*), he observed that 'to them of Java Pepper is cold'.

In the same series of notes Newton also remarked on the various locations of the brain that philosophers had invoked as the seat of the soul. He recorded various phenomena demonstrating that the brain could be badly damaged without affecting sensation. A frog would have its 'sence & motion' taken away if its brain was 'peirced', but a human would retain the use of his senses unless the piercing penetrated to the main blood vessels. A man could not, apparently, see through the hole that a trepan (or drill) made in his head, but 'the least weight upon a mans brain when hee is trepanned maketh him wholly devoyd of sensation & motion'.

A key element of his early research programme concerned the nature of free will, and the associated problem of how the soul was linked to the rest of the body. Some bodily motion was unconscious. Under the general heading 'Of Motion', Newton recorded that many human actions were purely mechanical: musicians could play without thinking, singers sing 'neither minding nor missing a note', and people walked without being conscious of how they did so. Vomiting induced by sticking a whalebone down one's throat was another example of an action that was purely mechanical, and it apparently proved the actions of animals to be 'mechanicall and independent of soules'.

Nevertheless, Newton's account of the soul involved a vigorous rebuttal of any purely mechanical explanation for its actions. Like most of his contemporaries he did not want to be tainted with the atheistic reputation of mechanical philosophers such as Descartes and Thomas Hobbes. As the faculty of the soul linked to personal identity, memory offered significant evidence relating to the springs of human action. Blows to the head could cause it to disappear completely, while it could be reactivated by similar events occurring much later. In an entry entitled 'Of the soule' he argued that memory consisted of more than the action of 'modified matter', and that there had to be a 'principle' within us that enabled us to call something to mind once the original action had ceased. This insight would be one of the cruxes of Newton's later natural philosophy.

In another extraordinary short essay entitled 'Of Creation', he discussed the 'souls' of animals, which most philosophers of his day believed were of a completely separate nature from those of humans. Newton suggested that there was a sort of primordial 'irrationall soule' which when joined to different kinds of animal bodies made all the various brutes that now existed. In shorthand (because of the daring nature of his argument), he suggested that to say that God initially made specific souls for specific species was to assert that he had done more work than he needed. The differences between species arose from their instincts, which depended on the make-up of their bodies. More radically still, he argued that human souls were basically alike, and that the differences between people arose merely from distinctions in their constitutions. In a short, separate entry on God, he noted that neither men nor beasts could be the result of 'fortuitous jumblings of attomes'. There would have been many useless parts, 'here a lumpe of flesh there a member too much some kinds of beasts might have had but one eye some more than two'.

The most stunning attempt to distinguish between the actions of the soul and body began with a series of notes on the nature of the 'imagination' (or 'fancy') and creativity. The former was a faculty of the soul that produced images such as those found in dreams and memory. Newton argued that the imagination was helped by viewing things 'in a right posture with the heeles upward', as well as by 'good aire fasting moderate wine'. However, it was ruined by 'drunkenesse, Gluttony, too much study, (whence & from extreame passion cometh madnesse), dizzinesse commotions of the spirits'. 'Meditation', Newton warned, heated the brain in some 'to distraction', and in others led to 'an akeing & dizzinesse'. It was possible to train the imagination to do new things, and from Joseph Glanvill's *Vanity of Dogmatizing* (1661), Newton noted a famous story of an Oxford scholar who had learnt mind control from gypsies 'by heitning his fansie & immagination'.

Some time later than his entry on the Oxford scholar, but immediately following it in the text, he recorded a series of his own experiments on imagination and vision. At some point in 1665, he undertook a series of dangerous experiments on his own sight that involved staring at the Sun for an extended period of time. These were reported as subjective experiences, but his detailed description of a series of trials indicated an objective detachment. After he had stared at the Sun for some time with one eye, he noted that all light-coloured objects appeared to be red, while dark objects looked blueish. At first glance white paper appeared red when looked at with the damaged eye, but the same paper looked green 'if I looked on it through a very little hole so that a little light could come to my eye'.

The experiment was by no means concluded, for when (as he thought) the motion of 'spirits' in his eye had died down, he could produce an after-image of the Sun by shutting his eye. There appeared a blue spot, which grew lighter in the middle, gradually being encompassed by concentric circles of red, yellow, green, blue, and purple. Varying the experiment under different conditions, he noted that the spot would sometimes turn red. When he opened his eye again, he would see colours in exactly the same way as after the initial experiment. He concluded that the Sun and his imagination had exactly the same manner of working on the spirits in his optic nerve and brain. Outside, he looked at a cloud and witnessed the same reddish effects ('onely for the most part blacker') as when he stared at the white paper, and after a while he could make a spot 'glitter amidst the dusky red' when he looked at a cloud that was so bright his eyes watered.

The fact that this only constituted the first of a series of such experiments says a great deal about Newton's uniquely intense dedication to his task. After giving his eye some respite, he waited until an hour before dusk *and repeated all of the previous experiment*. Now, when he looked with his good eye on white objects such as paper or clouds, he could see an image of the Sun against their background, the image being surrounded by 'a dusky red & blacknesse'. He found it almost impossible to avoid seeing a solar image, unless he tried hard to set his imagination on other tasks. When the image of the sun was just about bearable in either eye, he could envisage several shapes in the place where the sun had been, 'whence perhaps may be gathered that the tenderest sight argues the clearest fantasie of things visible'. He added: 'hence something of the nature of madnesse & dreames may be gathered'. Such was the enduring power of these trials, that Newton recounted them in detail to John Locke in 1691, and did so again to John Conduitt in 1726, telling him that he could still conjure up an image of the Sun if he put his mind to it.

A new theory of light and colours

Some time after the initial entry on colours, Newton recorded a series of experiments with prisms on a new page with the same heading. With these, he not only refuted the Aristotelian notion of light and colour, but he also challenged the treatments of the topic to be found in the recent work of Descartes, Boyle, and Hooke. The exact date at which he embarked on these investigations is unclear, but in later accounts, he placed the initial impetus for his research in his efforts to replicate Descartes's report of experiments with a prism in his *Dioptrique*. In this work, Descartes had argued that the colours produced by transmitting light through a prism on to a wall about 50cm away from the prism served to explain the processes involved in creating a rainbow. At some point, Newton acquired a prism in order to reproduce this 'celebrated phenomena of colours', but the earliest experimental entries in the 'Philosophical Questions' notebook refer to two instruments.

The very first comment in the new section on colours was a proposal to test whether a mixture of prismatic red and blue made white. Already he had criticized older theories that held colour to be a mixture of black and white, or which assumed that colours arose through the mixing of shadows with light. Elsewhere in the notebook, Newton had also subjected to criticism the notion that light was caused by pressure. This had to be false, for the pressure of the vortex bearing down on us would make us see a bright light all the time, while one would be able to see in the dark merely by running. Finally, he attacked wave theories of light on the grounds that light travelled in straight lines, whereas waves or 'pulses' through an aetherial medium would not. Early on, he became committed to the idea that light was composed of corpuscles, or globules, an assumption that ran directly counter to the 'pulse' view outlined in the recently published *Micrographia* of Robert Hooke.

The key observation was described in the third of a series, in which he examined a thread – one half coloured blue and the other red – through a prism. One half, he noted, 'shall appear higher than the other & not both in one direct line, by reason of unequall refractions in the 2 differing colours'. He explained this differential refrangibility in terms of the underlying speed of the light 'globules', assuming that the slower moving rays were refracted differently from the quicker, and that the blues and purples constituted the slower rays. He inferred that bodies appeared as red or yellow whenever the slower rays were absorbed, and were seen as blue, green, and purple whenever the faster rays were not reflected. This was the basis of his later, more sophisticated account of how colours arise in natural bodies in terms of their disposition to 'exhibit' certain sorts of rays. As slow or fast moving globules, coloured rays were permanent features of ordinary light – which was a complex mixture of them – and individual rays were *revealed* but *not produced* by prismatic refraction. This ran counter to the universally accepted notion that prismatic colours arose through 'modifications' caused by refraction, and threatened both Aristotelian and standard mechanistic explanations of light and colour.

Nor was his work at this point separate from his understanding of the way in which the eye contributed to the experience of colours, and he proceeded to undertake a series of ocular experiments every bit as damaging as the sun-gazing trials. He deformed his eye by violently pressing it on one side, thus producing a number of 'apparitions', and then noted that he made a 'very vivid impression' by 'puting a brasse plate betwixt my eye & the bone nigher to the midst of the tunica retina than I could put my finger'. Newton repeated the act on a number of occasions, trying it in the dark, and also with various degrees of pressure. Needless to say, no other individual of the period did anything like this.

Measuring refractions

Newton continued his optical experiments in a so-called 'chemical' notebook, in which he entered another essay called 'Of Colours'. This was a radically different undertaking, which began with an account of examining a bi-coloured thread through a prism, but which then listed a series of highly original experiments on reflection and refraction. Where contemporaries (who had not known of differential refrangibility) had at most projected refracted rays a metre or so, Newton showed that different coloured rays had different indexes of refraction by projecting refracted rays onto a wall about 7m (22 feet 4 inches) away. In a dark room he let sunlight in through a tiny hole in the curtains, finding that when refracted through a triangular prism, the rays produced an oblong and not a circular shape on the wall. As he had noted before, blue rays were refracted more than red, although he was also careful to note that redness and blueness were not intrinsic to rays but were how specific rays appeared to the eye. With exceptionally precise measurements, he now determined that differently coloured rays emerging from the prism had their own specific degrees of refraction, a fact that no one until then had noticed.

Later in the series of experiments, he described a more complex arrangement in which the rays emerging from the prism were further refracted through a second. Blue and red rays each suffered the same degree of refraction as they had done from the first prism, and Newton noted that individually coloured rays were not further modified into other colours when refracted through the second prism. Introducing a third prism and setting them all parallel, he allowed emerging rays from all the prisms to overlap with each other; as he noted, 'where the Reds, yellows, Greenes, blews, & Purples of the severall Prismes are blended together there appears a white'. With these experiments he now had the fundamental features of what was to be his mature theory of light and colour. Ignoring his account of globules, he argued that white light was not a basic entity that gave rise to colours by being 'modified'. Instead, it was composed of a number (Newton did not at this point specify how many) of different primary rays, *each of which had its own immutable index of refraction.*

Another significant observation was his analysis of thin coloured films, a phenomenon originally observed by Hooke. Examining a flat piece of glass through a lens, placed as close to the glass as possible, one could see concentric rings of different colours. By considering the radius of curvature of the lens, Newton went as far as measuring the film of air that existed between the concentric rings and the plate to nearly one hundred thousandth of an inch. He developed this analysis in about 1670 or 1671, producing results that appeared first in his 'Discourse of Observations' sent to the Royal Society at the end of 1675, and then later in his *Opticks* of 1704. His main discovery was that the thickness of the film at any point was proportional to the square of the diameter of each circle. In addition to this, the difficulty he and others experienced in trying to bring about contact between the two pieces of glass would later constitute central evidence for the existence of short-range repulsive forces.

The second essay 'Of colours' also demonstrated vividly that eye experiments remained a central part of his project. Having dispensed with a brass plate as a valid tool, he got hold of a 'bodkin', a sewing implement for making holes in fabric, and once more thrust it into the recess behind his eye 'as neare to the backside of my eye as I could'. As before, a number of circles appeared, and as he put it, they were 'plainest when I continued to rub my eye with the point of the bodkin, but if I held my eye & the bodkin still, though I continued to presse my eye with it,' the circles would 'grow faint & often disappeare until I renewed them by moving my eye or the bodkin'.

Later, Newton stated that his discovery of chromatic aberration had put an end to his efforts to improve the grinding of lenses for refracting telescopes. Descartes had suggested that a lens ground into either of two conic sections (hyperbola or ellipse) would produce the clear image that could not be obtained with a spherical lens (because of the sine law of refraction). Newton himself had spent many hours attempting to do the same, and had recorded his results in the Waste Book. But chromatic aberration rendered all such attempts redundant, as different colours would be refracted differently and could not be brought to make a sharp image. If refracting telescopes were out of the question (though Newton did not entirely give up the idea), then perhaps he could make one that

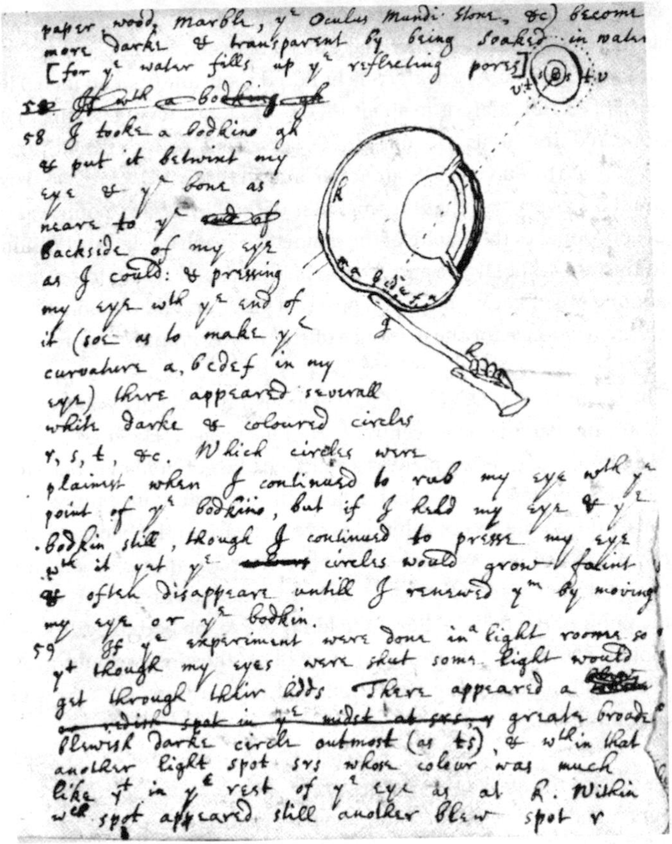

paper, wood, marble, y^e Oculus Mundi Stone, &c) become more darke & transparent by being soaked in water [for y^e water fills up y^e reflecting pores]

~~58 If wth a bodkin gh~~

58 I tooke a bodkin gh & put it betwixt my eye & ye bone as neare to y^e ~~end of~~ backside of my eye as I could: & pressing my eye wth ye end of it (soe as to make ye curvature a, bcdef in my eye) there appeared severall white darke & coloured circles r, s, t, &c. Which circles were plainest when I continued to rub my eye wth ye point of ye bodkin, but if I held my eye & ye bodkin still, though I continued to presse my eye wth it yet y^e ~~colours~~ circles would grow faint & often disappeare untill I renewed y^m by moving my eye or ye bodkin.

59 If ye experiment were done in a light roome so y^t though my eyes were shut some light would get through their lidds There appeared a ~~redish spot in ye midst at srs, a~~ greate broad blewish darke circle outmost (as ts), & wthin that another light spot srs whose colour was much like y^t in ye rest of ye eye as at k. Within wch spot appeared still another blew spot r

5. Newton's drawing of his deformation of his eye by means of a bodkin

used a mirror? Where contemporaries had merely discussed the theoretical possibility of constructing such an instrument, Newton went ahead and built a successful version, making every aspect of the device with his own hands. The metal easily tarnished and the image was devoid of colour, but it solved the problem of chromatic aberration and magnified as much as a good refractor. It was a remarkable achievement, and one for which Newton – reprising his Grantham role – became famous at Cambridge.

Chapter 4
The censorious multitude

The major turn that Newton's life took after he became a major fellow of the college in 1668 was to a large extent facilitated by Isaac Barrow, who had by now recognized Newton's potential. He thanked Newton (although not by name) for help in revising his *Eighteen lectures* on optical phenomena of 1669, and Newton almost certainly attended his Lucasian lectures on geometrical optics in 1667 and 1668. Barrow was presumably unaware of the radical nature of Newton's work in that area but with his support, Newton was elected as his successor in the Lucasian Chair in September 1669.

Early in 1669 Barrow had shown Newton a copy of Nicholas Mercator's *Logarithmotechnia*, published at the end of the previous year. Mercator had discovered a way of deriving values for logarithms by using infinite series; Newton claimed later that when he read the work, he had assumed (wrongly) that Mercator had uncovered the generalized binomial theorem for expanding polynomials with fractional powers. In any case, seeing Mercator's book and realizing that Mercator had begun to 'square' terms to produce infinite series prompted him to compose a remarkable mathematical tour de force, now known as 'On analysis by infinite series' (or 'De Analysi'). He did not specify the binomial theorem in this work but, amongst other treasures, laid out a number of infinite series that approximated to values for sin x and cos x, along with techniques for integrating the cycloid and the quadratrix. He announced that the methods of tangents and quadratures were inverse techniques, and drew from the October 1666 tract to offer a powerful basis for his method of fluxions. He would draw from 'On analysis' in two major mathematical letters written to Leibniz in 1676.

Barrow communicated this work to the London mathematician John Collins at the end of July 1669, revealing Newton's authorial identity a month later. Infinite series were all the rage, and via Collins, Newton's achievements, as well as the actual text, came to the attention of other mathematicians. In fact, in November Newton met Collins in London, where they discussed his reflecting telescope, series expansions, harmonic ratios, and the fact that Newton ground his own lenses. However, Collins noted that he was unwilling to disclose the general method underlying his work. At this time Barrow asked Newton to comment on the *Algebra* of Gerard Kinckhuysen, which Collins had recently translated. Newton's extensive remarks were never published but in any case he exhibited what Collins thought was a bizarre unwillingness for his name to be attached to the piece. He made it clear to Collins in September 1671 that he wanted his work to appear anonymously – if it appeared at all – and he had no desire 'to gain the esteeme of one ambitious among the croud to have my scribbles printed'. This attitude would govern his relations with potential audiences for his work for the next three decades.

Newton's Lucasian lectures on geometrical optics differed dramatically from those given by his predecessor. He employed a barrage of experiments, prisms, and lenses to corroborate his theory of the heterogeneity of white light and placed a major emphasis on the mathematical precision and certainty that attended his work, urging that natural philosophers should become geometers and should stop dealing with knowledge that was merely 'probable'. Here was Newton's first public pronouncement that natural philosophy could reach an absolute level of certainty and should be based on mathematical principles.

At this point Newton could have published work that would have stamped him as one of the most fertile scientists, and certainly the most brilliant mathematician the world had seen. Collins spent some time pushing him into publishing both 'On analysis' and a version of his optical lectures, and Newton expended a great deal of effort revising them, expanding the first (in early 1671) into a new treatise on methods of series and fluxions. He also rewrote his optical lectures in the second half of 1671, producing a new version that differed from the earlier one in that it suggested that one should measure refractions and reflections before discussing the nature of colours. However, when Collins prompted him again in April 1672, Newton told him that he had been thinking of preparing a joint publication of his optical and mathematical work, but had desisted, 'finding already by that little use I have made of the Presse, that I shall not enjoy my former serene liberty till I have done with it'. Nevertheless, at this time his name did appear as the editor of a book on geography by Bernard Varenius, a work to which he later admitted he had added little.

The cause of Newton's disillusionment was his first contact with an international audience. Collins had already been informed by Newton of the existence of his reflector, and the topic was 'live' again at the end of 1671 when Barrow delivered a new version of the instrument to the Royal Society. It was much admired by the fellows, and was examined in some detail 'by some of the most eminent in Opticall science and practise', as the Secretary of the Society, Henry Oldenburg told him. Oldenburg told Newton that a description of the instrument's construction and capacity had been sent to Christiaan Huygens at Paris, 'to prevent the arrogation of such strangers, as may perhaps have seen it here, or even with you in Cambridge'.

In reply Newton adopted his standard aloofness about his own invention, telling Oldenburg he had had the device in Cambridge for some years without making any great song and dance about it. He added advice on how to produce an alloy for the mirror and

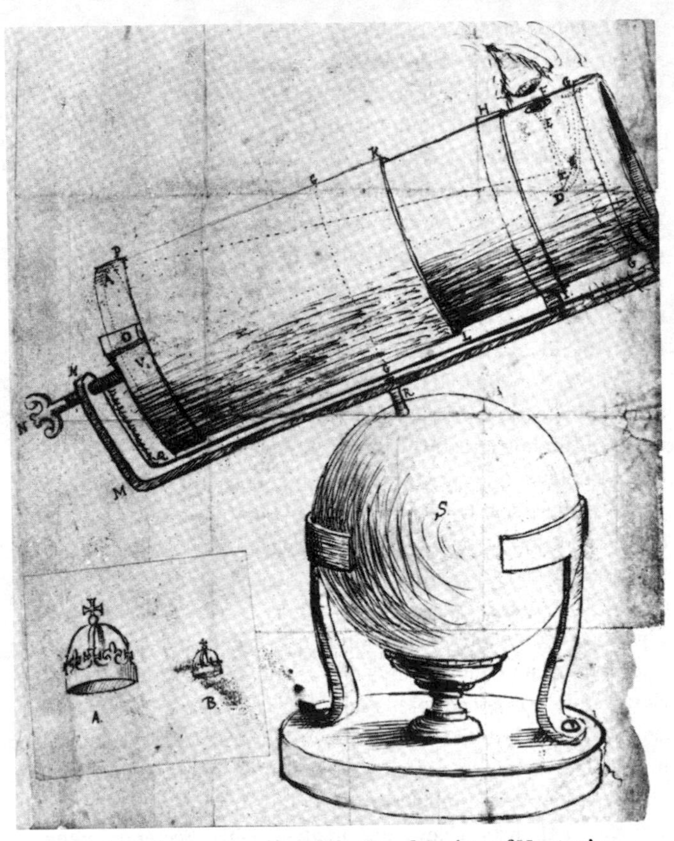

6. A sketch made by a member of the Royal Society of Newton's reflecting telescope presented to them by Isaac Barrow at the end of 1671

thanked the Society for electing him a fellow. He continued his pose of modesty in accepting the offer of a fellowship of the Society, offering to convey to them whatever his 'poore & solitary endeavours' could do to benefit their activities. Nevertheless, a further letter revealed that he had been prompted into constructing the reflector by what was in his judgement 'the oddest if not the most considerable detection which hath hitherto been made in

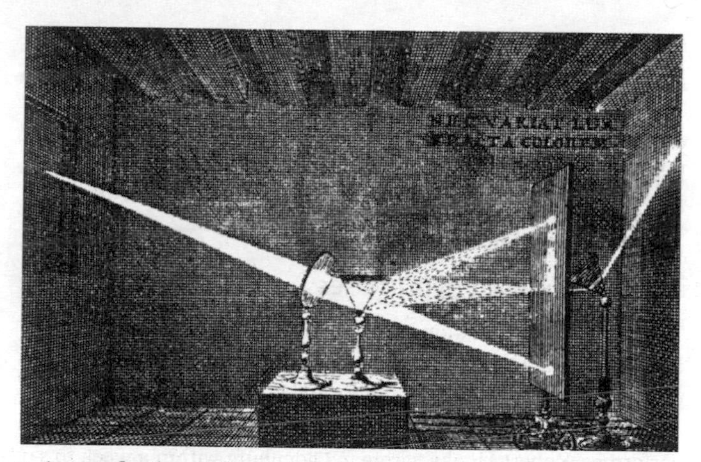

7. A reproduction of the crucial experiment, from the 2nd French edition of Newton's *Opticks*

because of their tendency to reflect certain rays and not others. He concluded by saying that it was much more difficult to determine what light actually was, or how it was refracted, or 'by what modes or actions it produceth in our minds the Phantasms of Colours', although he offered a hostage to fortune by asserting that it could *perhaps* no longer be denied that light was corporeal (i.e. made up of bodies). However, the last claim was not essential to his argument, he said, and he would not 'mingle conjectures with certainties'.

The essay was not merely the most radical challenge to accepted views about optics in modern history, but was a clear statement about what Newton took to be the proper way to investigate and justify scientific claims. In reply, Oldenburg remarked that the fellows had considered the paper with 'a singular attention and an uncommon applause', and had asked for it to be printed in the *Philosophical Transactions*. He also mentioned that the Society had decided that some of its members should attempt to repeat the experiments described in the paper, as well as some other relevant ones. Newton replied that he had sent his paper to the Society on account of their being the 'most candid & able Judges in philosophicall matters', and remarked that he deemed it a 'great 'privilege that instead of exposing discourses to a prejudic't & censorious multitude (by which many truths have been baffled & lost)', he could now 'with freedom' turn his attention 'to so judicious & impartiall an Assembly'.

The trouble with hypotheses

The combined publication of the description of the telescope and the paper on light and colours made him famous. A number of contemporary philosophers, most notably Christiaan Huygens, expressed their approval. However, the Royal Society's star performer, Robert Hooke, wrote to Oldenburg within a week to say that he had grave reservations about the theory. Although he agreed that the phenomenon was true, he did not believe that differential refrangibility could only be explained by Newton's theory of the heterogeneity of white light, nor did he agree that it showed that light was corporeal. Hooke announced that he had found similar effects before, and he could not agree that Newton's theory of white light was as certain as Newton made it out to be.

Hooke's own hypothesis, namely that light was a pulse or motion transmitted through an undifferentiated and invisible medium – with colour being a modification of light caused by refraction – was, he asserted, based on hundreds of experiments. If Newton really did have a single compelling crucial experiment that proved his own thesis, then Hooke would readily concur with Newton's theory. However, he could think of numerous other hypotheses that would also explain what had happened. Why should all the motions that make up colour be in the white light *before* it hit the prism? There was no necessity for this to be the case, any more than there was that the sounds were 'in' the bellows that later issued from the pipes of an organ. Newton's theory was merely a hypothesis, if a 'very subtill and ingenious one', and not nearly so certain as a mathematical demonstration.

Newton's lengthy response to Hooke of June 1672 used a wealth of data from his optical lectures as well as from his laboratory notebook, and in itself was a major contribution to optics. The reply started with a haughty rebuke about Hooke's behaviour. He should have 'obliged' Newton with a private letter, while the 'hypothesis' Hooke had ascribed to him was not the one Newton had expressed in his paper – for nothing hung on whether light were a body or not. Ignoring 'hypotheses', which he despised, Newton had spoken of light 'in generall termes, considering it abstractedly as something or other propagated every way in straight lines from luminous bodies, without determining what that thing is'.

Newton then launched a direct assault on Hooke's wave theory of light, using arguments that he had developed while a student. One might accept, he said, that Hooke's hypothesis could account for the phenomena Newton had described, but it was beset with difficulties. Waves and vibrations of fluids did not travel in straight lines, as rays of light seemed to do; worse, given that different bodies would necessarily exude 'unequall' pulses, then ordinary light must be a mixture of these unequal pulses, or 'an aggregate of difform rays', which was the very sort of heterogeneity for which Newton had argued. Newton strode on: Hooke's hypothesis was not only insufficient but unintelligible, and if he were a half-decent experimenter, he would have found that what Newton had said was true. Considering light 'in general', there were more than two basic colours, contrary to Hooke's claim, while the crucial experiment was indeed as Newton had described it.

Hooke could scarcely mistake the tone. In a letter to a senior member of the Royal Society he noted that he had since performed further experiments with prisms and coloured rings, as Newton had suggested, but remained unconvinced by Newton's theory. Nevertheless, he added that he was sorry if Newton had been offended by what he had written, since it had never been meant for his sight. Hooke stressed that he did have good evidence for his views, and indeed he had produced diffraction experiments which showed that in certain conditions light really did spray out into 'shadowed' areas. He was sorry if his own hypotheses were unintelligible, although he sarcastically noted that he made 'noe question' that Newton could explain how primary rays maintained their own constant refrangibility after refraction, and were then made to converge again 'and unite into one and then every one part againe and keep on their way Direct & undisturbed as if they had never mett'. Newton might understand this but Hooke did not, and nor did he understand why Newton was now afraid of saying what a ray of light actually *was*.

Hooke's response, wedded as he was to a view that philosophical explanations had to refer to intelligible physical causes, set the pattern for the way natural philosophers would understand Newton's programme. Early in the following year Christiaan Huygens repeated the point made by Hooke to the effect that there were a limited number of basic colours from which all the others could be made. He also stated that Newton had not abided by the fundamental tenet of the mechanical philosophy, namely that he was obliged to devise a physical hypothesis that would account for the different prismatic colours. Until he had done this, Huygens remarked, 'he hath not taught us, wherein consists the nature and difference of colours, but only this accident (which is very considerable,) of their different refrangibility'.

This seems to have been the last straw for Newton, who told Oldenburg that he wanted to resign from the Royal Society, being unable to benefit them on account of his 'distance' from London. At the same time, he told Collins that he had experienced some 'rudeness' from members of the Society, a remark which got back to Oldenburg. With reference to Hooke, the Secretary of the Society told Newton that every group had a troublemaker, and 'that the Body in general esteems and loves you'.

Nevertheless, Newton did send an intemperate reply to Huygens. It was impossible, he said, to concoct the prismatic colours from yellow and blue, and it was inconceivable how the basic phenomena of light could be caused by only two sorts of rays. Although he had mentioned the fact in his original paper, Huygens's comment forced Newton to reiterate that simple and compound rays might look identical, and could only be distinguished by experiment. If white light *could* be made from two coloured rays alone, then it meant that the rays were already compound and not primitive. As if the tone were not sufficiently rude, Newton ranted that this was so obvious 'that I conceive there can be no further scruple especially to them who know how to examin whether a colour be simple or compounded & of what colours it is compounded'.

Despite being told by Oldenburg that Newton was a man of great candour, Huygens was irritated by Newton's attitude, commenting that he did not want to dispute with Newton if he defended his theory with such heat. Nevertheless he generously sent Newton a copy of his extraordinary *Horologium Oscillatorium*. Newton thanked him for his book, which was full of 'very subtile & usefull speculations' (such as the equation for centrifugal force), but he responded to the criticism of his tone by saying that it had seemed 'ungratefull' to him to have met with objections that he had already answered. To Oldenburg, in a letter that contained his response to Huygens, he repeated his intention to be 'no further sollicitous about matters of Philosophy'.

Newton continued to correspond intermittently with Collins and other mathematicians, discussing short-cut techniques for facilitating the construction of tables of logarithms, square numbers, and square and cube roots. However, other issues had by now crowded into his life. In late 1674 he was faced with the need to be ordained and hence to affirm his commitment to the Holy Trinity in order to retain his fellowship. For reasons explained in the next chapter, this was no longer possible and in January of the following year, he implied to Oldenburg that he was about to lose his position at Trinity. Nevertheless, after a trip to London at the end of February to meet high-ranking government officials, Newton received special dispensation to continue as a fellow without taking holy orders in the spring of 1675. The support of Barrow (now master of the college) may well been central to his success.

Cloudy days and bad prisms

Just as Newton thought he was free from disputing in the public arena, a new rash of correspondence pulled him back in. A critique by the Liège Jesuit Francis Linus opened up a new sort of attack on Newton's theory, which was continued by colleagues on his behalf when Linus died in 1675. This concerned the practical difficulty of following in detail the various instructions that Newton had given in his papers, and of achieving the outcomes that he said would ensue. To some extent Newton had foreseen such problems, which were inherent in his mathematicist approach, dealing as it did with one or two abstract and idealized experimental situations rather than a set of detailed descriptions of many related experiments. When Oldenburg sent him some queries written by fellows of the Royal Society in response to his initial paper on light and colours, Newton had admitted that his exposition had been obscure, and that his descriptions might have been longer and contained more diagrams if he had intended them for publication.

The trouble with Linus was magnified in ensuing correspondence with the Jesuit's colleagues, John Gascoignes and Anthony Lucas. Although a number of British natural philosophers appear to have repeated most of Newton's experiments without much trouble, the correspondence with the Jesuits proved how difficult it was for some highly accomplished philosophers to reproduce his experiments, or even to understand what their point was. For their part, the Jesuits believed they were following the tenets of the Royal Society in holding that scientific knowledge could only be built up gradually, by producing a number of different experiments that shed light on different aspects of the theory. Since it was so novel, they said, it was up to Newton to prove his theory. Newton, who felt that the Jesuits were explicitly attacking his sincerity and competence, argued that his crucial experiment *alone* was enough to make good his theory. He criticized the Jesuits for not following his instructions, for being unable to measure refractions to the required degree of precision (minutes and not merely degrees), for using inadequate prisms and for relying on long-dead witnesses to experimental trials.

At some point in 1677 Newton decided once more to publish his own optical work (probably in conjunction with his work on infinite series), consisting of a mixture of his optical lectures and his published correspondence. He was engraved for a frontispiece by the artist David Loggan in March of that year but things did not go according to plan. In February 1678 Newton asked Lucas for a copy of an earlier letter (of October 1676) sent by the Jesuit, which Newton had lost in a fire that must have only recently destroyed many of his papers and which put paid to the projected work. By chance, Lucas had already received permission from Newton two months earlier to have the same letter published in the *Transactions*, and it was passed on to Newton by Robert Hooke, one of the new secretaries of the Society following Oldenburg's recent death. However, Newton somehow realized that the version Lucas had sent Hooke for the *Transactions* was slightly different from the original. In one final letter to Lucas of March 1678, Newton spewed a torrent of abuse over the quality of science represented in Lucas's earlier letters. On the verge of a breakdown, he described Lucas and his 'friends' as comprising a Jesuit conspiracy against him. They had 'pressed' him into public disputes, the very thing that Newton hated most. Newton told Lucas that he had thought most of his points too 'weak' to acknowledge while there was 'other prudential reasons' why Newton was unwilling to 'contend' with him. Yet if public disputing with Jesuits was deeply unpalatable, there were other interests to take up his time.

Chapter 5
A true hermetic philosopher

Alchemy enjoyed a chequered reputation by the middle of the 17th century. Although many despised it as the hopeless quest to turn base metals into gold, others thought it had a long and venerable tradition, its secrets all the more significant and 'noble' for being obscured in recondite language and imagery. Natural philosophers such as Robert Boyle despised certain so-called alchemical practices while simultaneously believing that, if properly understood, some alchemical texts offered accounts of the most valuable operations in nature. As such, alchemy was part of a larger practice that was termed 'chymistry'. This included ordinary or 'vulgar' chemical operations that were part of any chemist's repertoire, but the alchemical tradition, which held all nature to be alive, seemed to promise answers to questions concerning fermentation, heat, and putrefaction, as well as the growth of animals, plants, and minerals. Alchemists were supposed to have access to techniques that mimicked these extraordinary processes and that would allow them to transmute various elements into each other. Most alchemists believed that there was a fundamental religious or spiritual aspect to the art, though evidence for this is conspicuously lacking in Newton's alchemical papers.

In the 1650s a circle of practitioners had developed in London around the American George Starkey, who developed the vitalistic theories of J. B. Van Helmont in a number of works. Newton turned to Starkey's work at the end of the 1660s, enticed by his analysis of the way that certain primary elements could be made to ferment or vegetate. In notes from this period, he also created a chemical dictionary from terms he found in Boyle's writings and noted down all the chemicals, procedures, and many pieces of equipment that were required to pursue the art of ordinary chemistry. However, he turned to the alchemical tradition to provide him with answers to questions concerning the most significant subjects that perplexed him and his contemporaries, namely fermentation, transmutation, life, reproduction, and the mind–body relationship. The exact date at which Newton became committed to the study of alchemy is not known, though a letter to his friend Francis Aston, and his purchase of two furnaces and Lazarus Zetzner's six-volume *Theatrum Chemicum* – all in 1669 – suggests that his elevation to the Lucasian Chair at the same time may have been a distraction from a deeper interest.

Nevertheless, Newton by no means neglected what 'common' chemistry had to offer. At about this time, in his 'chemical' notebook, he took many pages of notes from Robert Boyle's *New Experiments and Observations touching Cold* of 1665, adding queries and occasional experiments of his own. Along with other of Boyle's works published in the 1660s and 1670s, this represented the greatest mine of information that Newton had at his disposal, and gave detailed and authoritative information about the natural world. He noted, for example, that, despite it being much colder in Asia, the Chinese did not feel the cold as Europeans did, on the grounds that there were 'subterraneous exhalations' that contained 'calorifick streams'. Other notes from Boyle, along with Newton's musings upon them, reflected perennial topics of interest within his natural philosophy, such as heat, light, transmutation, and the 'principles' of nature.

Elsewhere, in a section on the transmutation of 'forms', Newton noted (from Boyle's *Origine of Formes and Qualities*, of 1666) that various living substances such as corals, crabs, and crawfish would turn to stone when pulled out of water, that near Sumatra there grew twigs that had 'worms' as their root, and that in Brazil, an animal akin to a grasshopper turned into a vegetable. Again, Boyle supplied crucial information concerning a white, earthy residue that remained when rainwater was distilled; Newton remarked approvingly that Van Helmont thought water was the principle of everything because all things 'by successive operations' could be 'reduced' to it.

The excerpts from George Starkey's *Pyrotechny Asserted* that follow the reference to Van Helmont indicate a shift of emphasis in his reading. At about the same time he read and took notes on Michael Maier's *Symbola aureae mensae duodecim*, which along with a manuscript on advice for travellers formed the basis for the letter to Aston in May 1669. This began with pompous advice on how to deal with foreigners, but Newton also told Aston (about to embark on a tour of Europe) to be on the look out for transmutations from one metal to another, for they would be 'worth your noting being the most luciferous & many times lucriferous experiments too in Philosophy'. Specific instructions came from his reading of Maier, but were also related to information gleaned from his reading of Boyle. Newton soon began to devour a treasury of both manuscript and printed alchemical works. Of great significance is the fact that his notes come from manuscript versions of many of these texts, indicating his acquaintance with a circle of alchemists based in Cambridge or more probably London. Unfortunately, the identity of many of these characters is elusive.

The vegetation of metals

As in the case of the 'Philosophical Questions', he quickly began to perform novel experiments, although indexing and comparing different alchemical works and terms would remain a key element of his research strategy. Soon after he bought the *Theatrum*, he created a short list of 'Propositions' in which he drew from texts cited in the collection. Here he referred to an active, 'mercurial spirit', termed 'magnesia', which was the 'unique vital agent' that permeated all things in the world. This was the so-called philosophical mercury – the primordial form of all metals that when reconstituted by alchemists could perform extraordinary transmutating effects. Working by means of a gentle heat, it could be harnessed to reduce (or 'putrefy') elements into their most basic state, and then to 'revivify' (or 'generate') them into a new form. Invoking an analogy that was basic to the alchemical tradition, Newton remarked that the *modus operandi* of this spirit was specific to whichever realm it worked in, whether on metals, or in the human body, or in the alchemist's laboratory. From 'metallic semen' it would generate gold, while from human semen it could generate human beings.

Similar themes were treated at much greater length in a remarkable paper that was probably written in the early 1670s. Known by its first line, 'Of Natures obvious laws and processes in vegetation', this

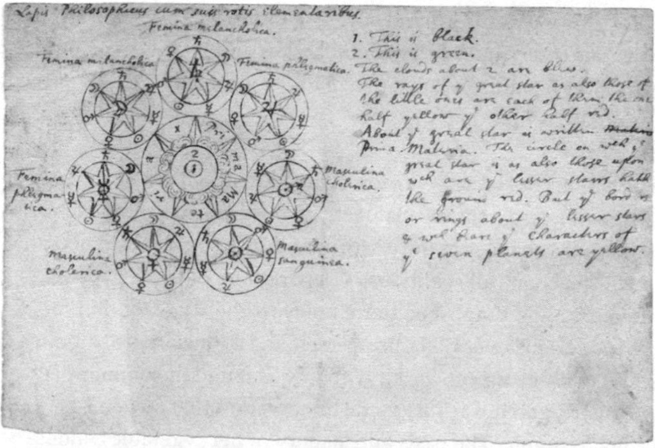

8. Newton's own drawing of the Philosopher's Stone; this was a substance that could help perform alchemical operations such as turning base metals into gold, or rejuvenating human beings so as to grant them immortality

is among the handful of Newton's most important writings on any subject. Many of the themes – transmutation, condensation, nature as a perpetual 'circulatory worker' – would appear again and again in different areas of his work. Under numbered headings resembling an outline of a treatise, Newton noted that the same laws that governed vegetable growth also covered the development of metals. By means of the alchemical art, metals could be made to vegetate as a result of working the 'latent spirit' that lay in them. As in art, so Nature could only nourish, but not create the various 'protoplasts' or forms of natural things – the latter being God's work.

There were a number of areas of 'agreement' between the different kingdoms of animals and metals, and, as we have seen, for Newton growing metals in laboratory conditions was analogous to the way Nature did her work. Indeed, it was because metals were living things that they had a tremendous capacity to act on animals for better or worse. This was visible from the rejuvenating power of springs, the fact that variation in the amount and type of metallic particles in the air gave rise to 'healthfull & sickly yeares', and in the observation that the ground covering mines was often barren. Minerals could unite with animal bodies and become part of them, 'which they could not doe if they had not a principle of vegetation in them'.

The actions of Nature, Newton claimed, were either 'vegetable' (or 'seminall'), or purely 'mechanicall', as in so-called 'vulgar' chemistry. Sometimes changes in the textures of things could be brought about by the 'mechanicall coalitions or separations' of these particles, but often it was accomplished more nobly by the action of the 'latent vegetable substances'. Nature herself had a much more 'subtile secret & noble way of working' than was found in common chemistry, and it was this mode of operation that alchemists were trying to mimic. The basis and 'agents' of her vegetable actions were the 'seeds or seminall vessels' in the heart of matter (surrounded by a humid covering), which Newton called the 'fire', 'soule', and 'life' of Nature. These constituted an 'unimaginably small portion' of matter, activating what would otherwise be only a 'mixture of dead earth & insipid water'. Whereas the body of grosser matter was often unaffected by extreme heat, the 'virtue' of the seeds could be terminated or corrupted by even a marginally excessive rise or fall in temperature. Vegetation was a central part of the process, consisting in the action of 'mature' seeds on the less mature parts of a different substance to make it as mature as itself. In another part of the manuscript he referred to the subtle part of matter as the 'vegetable spirit', which was the same in all things save for the degree to which it was 'digested' or mature. When different vegetable spirits were mixed they 'fell to work', putrefied, and 'mixed radically & so proceed in perpetuall working till they arrive at the state of the less digested'.

The paper also dealt with the ways in which, by imitating nature, alchemists could make use of vegetation to effect extraordinary phenomena. An initial stage of reduction to a 'putrefyed Chaos' was requisite for the alchemist's work to take place and 'all putrefyed matter [was] capable of having something generated out of it'. Putrefaction 'alienated' something from what it was, and was the condition for generation and nourishment, although total putrefaction made 'a black stinking rottenness'. As in Nature, this was to take place in a 'gentle' heat, and on moist substances, while coldness or extreme heat would ruin the work. Alchemy could promote the action of Nature on anything whatever, and the product was no less 'natural' than if Nature had produced it alone: 'Is the child artificiall because the mother took physick, or a tree less natural which is planted in a garden & watered then that which grows alone in the field?' By art, a 100-year-old oak could be made to propagate, and 'duly ordered and mixt with due minerall humidity', minerals could be made to 'rot & putrefy'.

The mineral cosmos

Newton argued that when metals were transformed into 'subtle & volatile fumes' they could pervade water (or other liquors) and 'impregnate it'. In cold water they lost their vegetating power, congealing into a 'fixed' salty state, and it became extremely difficult to turn them back into metals. Sea-salt, for example, was a mixture of different sorts of metallic fumes all concreted together, and these 'saline clusters' had a further ability to join up and form long crystalline tubes. This tendency of fumes or vapours to congeal was observed by distilling rainwater into its constituent parts, and also by noting the tendency of water to bind with minerals to form growths on rocks.

The most 'intimate' condensing was achieved by mixing the 'invisible vapors' of different sorts of fumes. These produced concretions of a more 'open & subtile constitution', such as 'nitre' – an element Newton called a 'spirit' that was the 'ferment of fire & blood . . . & all vegetables'. When they thickened, the humid parts that gave rise to nitre also created salt, but the cold of the sea stifled these more subtle vapours and so nitre was never found in it. In its more subtle nitrous state, salt would ferment and putrefy but gross salt was itself 'dead'. Either naturally or artificially, salt could however be 'incited' to vegetate 'by other substances that are in a live & vegetating state'. Basic salts preserved meats and worked by means of their gross particles but in certain circumstances their 'latent principle' could be triggered to work 'vigorously' on other elements. In this state nitre was generally held to be the most powerful mineral for enriching land, as well as being the source of gunpowder and of the purest part of air. If salts could be made to putrefy, Newton suggested, they could make a wonderful medicine.

Recalling his earlier theory, Newton described a great circulatory system in which various watery vapours and mineral fumes were exhaled upwards by the Earth. As air rises so aether is constantly forced downwards into the earth '& there its gradually condensed & interwoven with bodys it meets there & promotes their actions being a tender ferment'. Being sticky and elastic, it brings down heavier bodies as it descends, and being much finer than air, it does so with a much greater speed than that with which the air rises. The

Earth was thus like a great animal or 'inanimate vegetable', breathing in the aether for its daily refreshment. Terrestrial elements were composed of aether mixed together with a more active spirit, he continued, which was 'Natures universall agent, her secret fire, the ferment & principle of all vegetation'. The 'materiall soule of all matter', it could be activated by a gentle heat, and was perhaps in essence made up of or the same as light. Both had a 'prodigious active principle', both were 'perpetuall workers', and all things emitted light when heat was applied to them. Like sunlight, heat was necessary for generation, and 'noe substance soe indifferently, subtily and swiftly pervades all things as light & noe spirit searches bodys so subtily piercingly & quickly as the vegetable spirit'. This remarkably rich cosmology aimed at nothing less than uncovering the active elements of life and indeed the entire universe. In different guises, Newton would return repeatedly to the same themes.

Squashing tadpoles

Newton's explicitly alchemical cosmology appeared in a different form in a work composed late in 1675. Although it was not printed in his lifetime, his 'Hypothesis' of 1675 dealt with all the major topics that would reappear in the early 18th century as 'Queries' in the different editions of his *Opticks*. Whereas the alchemical work aimed more explicitly at uncovering the active elements in ordinary matter, the work described in his 'Hypothesis' was partly concerned with subjecting the aether to the same experimental forms of enquiry by which Robert Boyle had investigated the air. Indeed, when Newton met Boyle in the London in early 1675 the latter had apparently joked about Newton's intention to trepan the 'common aether'. At the same time Newton also had a lengthy discussion with Hooke about the cause of reflection and refraction, which Newton attributed to the action of the edge of the aetherial medium into which the light was passing. Repeating his proposal of a decade earlier, he told Hooke that an experiment in an air-pump could prove this, by showing that the phenomena of reflection and refraction would not be altered by taking place in an evacuated air-pump; it was the aether and not air that gave rise to reflection and refraction.

At the start of December 1675, Newton sent Oldenburg two pieces of work. One was the 'Discourse on Observations' mentioned earlier in connection with coloured rings, while the second was a short treatise Newton dismissed as 'another little scrible'. Read out at weekly meetings from 9 December, it was entitled 'An Hypothesis explaining the Properties of Light discoursed of in my severall Papers'. Although he had previously intended never to publish anything of this nature, he said (undoubtedly with reference to Hooke), 'I have observed the heads of great Virtuoso's to run much upon Hypotheses, as if my discourses [lacked] an Hypothesis to explain them by.' Optimistically, he said that he hoped this would put an end to disputes about his work.

According to this treatise, many types of terrestrial phenomena were caused by aether rather than air. Aether was more rare, subtle and 'elastic' than air, and was a compound mixture made up 'partly of the maine flegmatic body of aether [and] partly of other various "aetheriall Spirits"' in the same way that air was composed of the main body of air mixed with various 'vapours and exhalations'. The fact that the aether could give rise to such diverse phenomena as electricity and magnetism was ample proof of its compound nature. Perhaps, he conjectured, all of 'Nature' was composed of various amalgams of aetherial spirits or vapours that had been condensed by precipitation. The original forms of Nature were created by the immediate hand of God, and ever afterward by the power of Nature itself. By dint of the command 'Increase and Multiply', Newton continued (recalling the language of his alchemical tract), Nature 'became a complete Imitator of the copies sett her by the Protoplast'.

Newton described a simple experiment that shed light on the nature of electricity, and again invoked the notion of condensing. It involved the vigorous rubbing of a circular piece of brass-enclosed glass until miniscule bits of paper under the glass jumped up, 'mov[ing] nimbly to and fro'. The bits of paper would continue 'leaping' even after the rubbing had stopped, skipping and jumping in every direction while some rested on the underside of the glass for a short time. Evidently, he wrote, some 'subtle matter' in the glass had been rarefied and released from it, constituting a sort of aetherial wind. Afterwards it had recondensed and returned to the glass, thus causing the electrical attraction that drew the paper to its underside.

Ten years after his first crude musings on the subject, the 'Hypothesis' also gave Newton the opportunity to make public his thoughts on the possible causes of gravitation. This could be caused by the continuous condensing of some very refined 'gummy, tenacious & Springy' nature, analogous to the part of the air that supported life. This spirit might be condensed in fermenting or burning bodies and fall as gravitational rays into the Earth's cavities, forming 'a tender matter which may be as it were the succus nutritious of the earth or primary substance out of which things generable grow'. The Earth would then release an upward stream of aerial exhalations, which would ascend to the stratospheric layers of the atmosphere, when the matter would once again be 'attenuated into its first [aetherial] principle'. Again repeating the terms from the alchemical text, he noted that nature was thus a 'perpetuall circulatory worker', turning fluids into solids, refined matter into 'grosser' matter, and indeed all things into their opposites, and back again. On an even grander scale, the Sun might play a central role in exactly the same phenomenon, drinking up the aether 'to conserve his Shining' to prevent the planets from escaping.

For Newton aether accounted for most terrestrial phenomena, energizing activities like fermentation, putrefaction, melting, reflection, and refraction. More speculatively, he thought this might explain 'that puzleing problem', namely the capacity to move one's own body, with muscles contracting and dilating according to how one condensed or dilated the aether that pervaded them. Doubtless this theory was the basis of the discussion on trepanning the aether he had enjoyed with Boyle the previous spring, for Newton had proposed that Boyle attempt further air-pump experiments on muscles. Despite the fact that water could not be compressed, Boyle had managed to partly squash a tadpole, indicating that its 'animal juices', presumably with rarefied aether in tiny pores, could be contracted (and expanded). Newton even adapted Boyle's notion of the 'spring' or elasticity of the air to hypothesize that in normal situations there had to be a given amount of elastic or 'springy' aether inside a body in order to 'Susteyne & Counterpoyse' the pressure of the external aether.

At some point in the late 1670s or more likely in the early 1680s, Newton composed an extraordinary text ('On the gravity and equilibrium of fluids', now known as 'De Gravitatione') in which he argued vigorously against Descartes's notion that motion could only be measured relative to surrounding bodies. For our purposes, what is remarkable is Newton's contention that empty space was full of different *potential* shapes that were capable of 'containing' (but which were not, as Descartes would have it, the same as) physical objects of the same size. All space was an effect of, but not the same as God, who was able to make certain spaces impenetrable, or reflect light in a certain way, thus creating perceptible bodies – 'the product of a divine mind realised in a definite quantity of space'. According to Newton all this could be achieved by the mere action of divine thinking and willing, something that was analogous to the way in which we move our bodies at will. If the latter were known to us, he concluded, 'by like reasoning we should also know how God can move bodies'. Despite many differences between God and man, we were, after all, created in His image. As we shall see, a more grandiose version of this theory would appear in his major 18th-century writings.

By revealing how creatures controlled their own muscles, squashing tadpoles might therefore shed some light on the mind–body relationship. The way the soul controlled the relative densities of the fluids involved in muscular motion was tricky, but Newton bravely offered a number of hypotheses. Central to his view was his theory that the juices of animals contained aetherial 'animal spirits', which did not escape through the pores of the outer coatings of the brain, nerves, and muscles. The reason for this, Newton argued, was that certain parts of the body were more or less disposed to house this spirit in virtue of the fact that there was a 'secret principle' of aetherial 'sociability' or 'unsociability' between different substances. This allowed the spirit to remain in some parts of the body and not others, and the same theory might explain why the solar and planetary vortices remained separate. In the case of aerial particles, he suggested, a third element could be introduced to make previously 'unsociable' substances sociable to each other. Could the soul not interact with the aether in the same way by introducing a different form of aether that might render the animal spirits of the muscles and their coating sociable or unsociable to each other?

On Hooke's shoulders

Directly confronting the theory described in Hooke's *Micrographia*, Newton remarked in the 'Hypothesis' that light was neither the aether itself nor its vibrating motion, but 'something' – he would not say exactly what – that was exuded from lucid bodies. Some 'principle of motion' initially accelerated light away from these bodies, but again Newton would not say whether the cause of this were 'mechanical' or whether it was accomplished by some other means, possibly similar to the principle of self-motion that God had implanted in animals.

Light and aether acted upon each other, he continued, aether refracting light, and light acting on aether to make heat. Light could also cause aether to vibrate, sending vibrations cascading through a larger body in the same way that the beating of a pair of drums could stimulate the air to vibrate. By analogy with the way that vibrating air gave rise to sound, the experience of various colours could be caused by vibrations set up in the *capillamenta* of the optic nerve. The strongest vibrations would cause the most intense colours, and Newton even proposed that light could be analysed according to the way that sound was 'graduated' into tones. Indeed, it was in this paper that Newton first publicly suggested (on the basis of lines drawn by a friend) that the spectrum be divided up into seven colours, again on analogy with the octave. Finally he attempted to explain how concentric bands appeared in thin plates, and also how diffraction occurred. The latter had caused a disagreement at the Royal Society meeting in spring 1675 where Hooke had raised the topic, Newton asserting that it was merely a form of refraction, and Hooke affirming that, if so, it was a novel sort. In the 'Hypothesis', however, Newton now pointed out that he had read that, long before Hooke, Grimaldi had performed some diffraction experiments.

In a letter sent a week after he transmitted the 'Hypothesis', Newton described some further electrical experiments that could be tried with glass and bits of paper. These triggered a spate of attempted replications, and it is a mark of Newton's impact and originality that these offhand observations on electrical phenomena were to be deeply influential over the next four decades. Of more immediate concern to Newton was a growing dispute with Hooke over optical phenomena. At the reading of the second part of the 'Hypothesis' on 16 December, Hooke had stood up and remarked that the bulk of Newton's doctrine was contained in his *Micrographia*, and that Newton had merely carried it further 'in some particulars'. When Newton heard this he returned with interest the twin compliments of unoriginality and plagiarism to the Gresham professor. Hooke's account in *Micrographia* of the aetherial cause of optical phenomena differed little from those found in Descartes 'and others', Newton said, and he had 'borrowed' many of their doctrines, extending them further only by applying his version of the theory to the phenomena of thin plates and coloured bodies.

He had little in common with Hooke, Newton went on, save for the general notion that the aether vibrated – and then Hooke supposed light was identical with the vibrating aether, while he did not. He explained refraction and reflection, as well as the way in which colours of natural bodies were produced, very differently from Hooke, and indeed Newton's experiments on thin plates 'destroy all he has said about them'. This letter was read at a meeting of the Society on 30 December, and Hooke, undoubtedly taking umbrage at both Newton and Oldenburg, created a 'philosophical club' (containing allies such as Christopher Wren) two days later. Here he repeated the charge that Newton had effectively taken material and theories wholesale from the *Micrographia*.

When another letter from Newton was read out at the Royal Society on 20 January 1676, Hooke immediately dashed off a conciliatory letter to him, accusing Oldenburg of fomenting trouble between them. He knew what Newton wanted to hear, pleading that he detested contention and feuding in print, and protesting that he valued Newton's 'excellent Disquisitions'. Like other comments in the letter, his claim that he was pleased to see Newton 'promote and improve' notions that he had begun much earlier but had lacked time to complete was a double-edged sword. However, despite further comments to this effect, Hooke did lavish praise on Newton's abilities, which he said were greatly superior to his own. He ended by saying that he would be pleased to engage in a private correspondence with Newton, expressing his objections in personal letters if that was acceptable.

It was in this context that Newton composed his famous letter in which he said that, if he had seen further, it was because he stood on the shoulders of giants. Private correspondence was more like consultation, he told Hooke, and most welcome, since 'what's done before many witnesses is seldome without some further concern than for truth'. Inverting the stress of his letters to Oldenburg, he now praised what Hooke had done beyond Descartes, noting that it was even possible Hooke had performed experiments that he himself had not done. The last phrase, like many expressions hurled from both sides of this exchange, could be read in two ways, and whatever reconciliation there was would last barely four years.

A few months after this contretemps with Hooke had ended, Newton wrote to Oldenburg about a letter that had recently been published anonymously by Boyle on the subject of the alchemical mercury. This had heated molten gold when they were mixed together; although Newton suspected that the mercury may have operated on the gold by means of 'grosser' metallic particles, and therefore might not be of any use in medicinal or alchemical operations, he remarked that Boyle had done well not to publish more on the subject. Indeed, it might be 'an inlet to something more noble, not to be communicated without immense damage to the world if there should be any verity in the Hermetick writers'. Boyle should get the advice of 'a true hermetic Philosopher' whose judgement would be more valuable than that of anyone else – 'there being other things beside the transmutation of metals (if those great pretenders bragg not) which none but they understand'. Later, he would criticize Boyle for being too open and 'desirous of fame', a remark that surely refers in part to this episode.

An alchemical cosmogony

In February 1679 Newton wrote to Boyle regarding a discussion that they had conducted earlier on the notion of 'physical qualities', probably during his visit of spring 1675. Undoubtedly, this letter drew from his alchemical researches although it was also related to (and in many places is a summary of) the more conventional philosophical views he had expressed in the 'Hypothesis' of 1675. Newton told Boyle that there was an elastic aether diffused throughout the atmosphere and repeated his comments in the 'Hypothesis' to the effect that it could account for many standard phenomena. Once more he invoked his theory of 'sociability' to explain why some metals needed to be treated with a 'convenient mediator' in order to mix with water or other metals. Other parts of the letter drew from his alchemical work and he told Boyle that, considering how aerial substances were created by the continual fermentation of the bowels of the Earth, it was not so absurd to think that the most permanent part of the atmosphere was metallic. This was the 'true air', kept just above the ground and beneath the lighter vapours by the weight of its metallic particles. It was not the life-giving part of the air, however, and 'afforded living things no nourishment if deprived of the more tender exhalations & spirits that flote in it'. Newton's final flourish was a paragraph on gravitation, explained by invoking his aether theory.

For most of his career, Newton would be deeply committed – if for the most part only in private – both to aetherial hypotheses and to his alchemical programme. He experimented furiously in the late 1670s and early 1680s, and he returned to the topic as soon as he had finished composing the *Principia* in the spring of 1687. The bulk of his work consisted in organizing and assessing the quality of different texts, but another burst of experimental activity occurred in the early 1690s, when his friend Fatio de Duillier acted as an intermediary between Newton and some alchemists based in London. Active experimentation seems to have petered out when he went to London in the late 1690s, but he remained committed to investigating central themes within the alchemical tradition and indeed to the basic alchemical insight that nature was full of a subtle but powerful activity.

Occasional glimpses of his alchemical programme were revealed to others. In late 1680, when Newton was engaged in a protracted series of alchemical experiments, Thomas Burnet of Christ's College Cambridge went to Cambridge's best natural philosopher for advice on how God might have created the Earth through natural means. Burnet's *Telluris Theoria Sacra* ('Sacred Theory of the Earth') of 1681 would ultimately be the first work in the genre of physico-theology that became popular in the 1690s, by then based on the philosophy of Newton's *Principia Mathematica*.

Newton told Burnet that the creation of mountains and oceans might initially have been caused either by the heat of the sun, or by the pressure of the terrestrial and lunar vortices on the primordial waters. The earth would shrivel towards the equator, making the equatorial regions 'hollower' and thus allowing the water of the oceans to conglomerate there. Additionally, the first days would have lasted a lot longer than those of the modern period, giving the process of creation enough time to become approximately what it is today. To understand how the primordial chaos had become differentiated into hills and cavities, he returned to the analysis of his 'vegetation of metals' paper, in which he had noted that solids were often created in solutions, such as when saltpetre dissolved in water and crystallized into long bars of salt. Apart from this, the drying out and shrinking of other parts of the chaos under the heat of the Sun would leave channels for water to descend underground, and for 'subterranean vapours' such as geysers and 'fumes' in mines to rise from the depths.

In an important exercise in scriptural exegesis, Newton also criticized Burnet's account of how the Mosaic description of Creation should be understood. The account of the creation of two great lights (i.e. the Sun and the Moon) and stars on the fourth day was not supposed to imply that they were actually created on that day, nor did Moses describe their physical reality, 'some of them greater than this earth & perhaps habitable worlds, but only as they were lights to this earth'. Newton adopted a similar approach to the Mosaic description of the light that was created on the first day. Although Moses had 'accommodated' his language to the perceptual capacities of ignorant people, it was not thereby false. His description of Creation was not 'Philosophical or feigned', Newton argued, but *true*– 'his business being not to correct the vulgar notions in matters philosophical', but 'to adapt a description of the creation as handsomly as he could to the sense and capacity of the vulgar'.

Apart from the hint in the letter to Burnet, Newton also stated that the existence of other worlds was not implausible in a letter to

Richard Bentley in early 1693. The 'Philosophical Questions' notebook also indicates that as a student he already held the radical view that after a conflagration there would be a 'succession of worlds', and in 1694 he told David Gregory that comets had a special divine function, and that the satellites of Jupiter were held in reserve by the Creator for a new creation.

In an extraordinary conversation with John Conduitt at the end of his life, Newton told him that light and other material emitted by the Sun had coalesced into a moon and then into a planet by attracting other matter. Finally it had become a comet, which in time would fall back into the Sun to replenish it. He added that this comet might well be the same as the Great Comet of 1680, which would crash into the sun in the not too distant future. When it did so it would dramatically increase the Sun's heat to such an extent that 'this earth would be burnt & no animals in this earth could live', an event that seemed to explain the supernovas seen in 1572 and 1604. All this might be superintended by superior 'intelligent beings' under God's direction. Newton went on to say that human existence on the planet was limited and he implied that divine power might 'repeople' the planet. After this Conduitt pointed to a passage in the *Principia* where Newton referred to stars being replenished by comets and asked Newton why he did not make clear the implications for the future of our own solar system. Since the topic of the end of the world was evidently amusing, Newton remarked in a rare moment of levity that it 'concerned us more, & laughing added he had said enough for people to know his meaning'.

Chapter 6
One of God's chosen few

When Newton went to Trinity, he was introduced to a regime that placed great store by the study of writings of the Church Fathers, and of course, the Bible. At some point, probably in the early 1670s, he became a radical anti-trinitarian, holding that the conventional doctrine of the Holy Trinity was an incomprehensible and diabolical corruption introduced by perverters of scripture in the 4th century after Christ. Newton came to believe that the architect of orthodox Christianity, Athanasius, along with various monks, churchmen, and emperors of the Eastern and Western Empires, had polluted doctrine by introducing new words into Christianity, inserted false texts into the Bible and the writings of the Church Fathers, and packed church councils with their depraved supporters. At the heart of their project was the hideous view, as Newton saw it, that Christ was physically identical to God. Newton believed that he had been chosen by God to discover the truth about the decline of Christianity, and he believed it to be by far the most important work he would ever undertake.

Newton's need for special dispensation to be relieved from taking holy orders suggests that his heretical views had taken hold by late 1674. It is highly unlikely that he was invited to embrace these beliefs by anyone else, though like other undergraduates, under the principle of 'know your enemy', he was able to read similar views in contemporary anti-trinitarian writings. Nevertheless, anti-trinitarianism was deemed a terrible heresy by orthodox Christians, and there were severe punishments on the statute books for those who downgraded the nature of Christ. With the exception of two or three known sympathizers, Newton's entire life would be spent hiding his religious views from others.

Many of Newton's early notes betray an easy anti-Catholicism that would have been de rigueur for Cambridge students. If this was acceptable, Newton's downplaying of Christ in respect of God was not. Early on, he came to believe that there was ample scriptural evidence that Christ was different from and inferior to the Father, while pro-trinitarian texts were corrupt insertions or 'strained' misreadings. In many places in scripture Christ, the created Word or *logos*, admitted that he was a lesser being than God. If Christ had divine powers, and Newton thought he did, it was because God had allowed this to happen. God had permitted his Son to humble himself on the Cross, and indeed this made him worthy of being worshipped – but not as God. Christ had become the Son when the Word became flesh in the womb of the Virgin; it was this being alone, and not a human soul coexisting with a divine *logos*, that had suffered on the Cross. Finally, it was through God's will that Christ had been resurrected.

There's only one Whore of Babylon

For Newton, trinitarian doctrine was incomprehensible and false, defended by obstruse metaphysical arguments and imposed on heathens either by force or by diluting it with pagan practices. He placed great store by the simplicity of the basic tenets of Christianity, and stressed that only a very few beliefs about Christ – what Paul called milk for babes – were necessary for a saving faith. These were that Jesus was the Messiah predicted in the Old Testament, that he was the Son of God who was resurrected after humbling himself before his Father on the Cross, and that he would one day return to judge 'the quick and the dead raised to life'. Nevertheless, there were deeper truths in scripture, or 'meat for men', to be acquired by those 'of a full age' after being baptized and admitted into communion. This knowledge, to be acquired through protracted study, was of things that were not necessary to the Christian faith, and Christians were not to engage in disputes about them lest they lead to schism.

The most important object of study was prophecy, especially in the Book of Revelation, the last book of the New Testament. Newton agreed with many of the most significant 17th-century Protestant exegetes about the core techniques that were required for understanding Revelation. Like them, he believed that the images therein referred to a battle between good and evil that had kicked off at the end of the 4th century. Key symbols and descriptions referred to specific periods when the true church was persecuted and the enemies of truth held sway or were conquered by the righteous. Indeed, certain approaches were so standard that – as in the case of the work of his Cambridge precursor, Joseph Mede – he held that he was building on their foundational 'discoveries'. True to his own method, Newton was apparently able to discuss technical prophetic issues with at least one contemporary (Henry More) without revealing what this implied about the history of Christianity.

In one gigantic exposition of Revelation that is almost certainly from the period 1675–85, Newton provided 'demonstrations' of his views much as he would do in the *Principia*. He began by claiming that he had 'by the grace of God' obtained knowledge in the prophetic writings, and now that the time was at hand when they were to be revealed, he was duty bound to teach their meaning for the edification of the church. This did not consist of all Christians, but

> a remnant, a few scattered persons which God hath chosen, such as without being blinded led by interest, education, or humane authorities, can set themselves sincerely & earnestly to search after truth.

Now, searching scripture was a 'duty of the greatest moment', and failure to correctly discern the signs of Christ's Second Coming would leave Christians open to as much criticism as the Jews had received for failing to realize that Jesus was their Messiah. This was a task that could only be carried out by the pure of heart, and few were ready. The true faithful would also be identified by appearing to be despicable, while the 'reproaches of the world', Newton commented, were the mark of the true church.

Central to the process of interpreting the Bible was 'methodising' prophecy according to a set of rules. Many of these were standard elements within Protestant scriptural exegesis, such as the need to insist on only one meaning of a given place in scripture unless there were reasons for doing otherwise. In the first instance this would likely be a 'literal' sense but occasionally a 'mystical' one could be allowed. As for the latter, this had to be done according to the tradition of a prophetic 'figurative language' that had been observed by ancient interpreters. Turning without such a basis to a mystical reading of a passage was a delusion, and it was such licentiousness in interpretation that had given rise to every heresy Newton could think of. Interpretations had to be 'natural', and they had to reduce scripture to the greatest 'simplicity'. Most importantly, prophetic visions and images had to be harmonized with each other according to these rules before they were applied to historical events. The Apocalypse was hard to understand, but properly decoded it was of immense importance to the true church. The true religion could not be proved like a demonstration in Euclid, and would not convince more than a handful of people – but this was as it should be. It was enough, Newton concluded, 'that it is able to move the assent of those which he hath chosen'.

In accordance with his plan Newton wrote out a long list of prophetic 'definitions', which drew upon a number of different sources. In the prophetic 'style', the Sun referred to a king, the Moon to his next of command, and Stars to the great men of the kingdom. The Earth referred to the nations of the Earth, or the common people of a nation, while the Sea also referred to a people or to nations; together, the Earth and the Sea referred to two different sorts of people. Sometimes words could mean more than one thing, so that a mountain could refer to a city or a temple, depending on the context.

Having listed the definitions, Newton next showed how particular visions related to each other. While some images in Revelation were 'successive', i.e. referred to later or earlier events, others were held to be 'synchronal', i.e. they referred to different aspects of the same period. However, as we have seen, their connections could be displayed before relating them to specific events. Virtually all interpreters, Newton included, understood that the vision of the seven seals of the book that was shown to John at the start of the prophecy referred to successive events. The first six seals referred to a period before the Great Apostasy took hold. In the fifth seal, for example, descriptions of a Woman in childbirth and a persecuting red Dragon (Satan) ready to devour the child depicted the prospective fate of the true Church (the Woman) and the great danger faced by her offspring (the 'Manchild').

Soon after the opening of the seventh seal there arose from the Earth a Beast, which according to Revelation had two horns while speaking as a Dragon, which caused all men to receive the name of the number of the Beast (666) on their foreheads. The godly were depicted in Revelation by the 144,000 Elect who received the mark of God, and who were sealed up by an angel. Another image depicted the Lamb (Christ) standing on Mount Sion with the Elect, who had the name of God on their foreheads. The Dragon spewed a torrent of water from his mouth (by which Newton presumed was meant multitudes of corrupt people usually depicted by the Sea) while the persecuted Woman (the true Church) now attempted to fly into the wilderness, a process in which she is helped by the 'Earth', that is, the godly. After a short period, most interpreters now turned to the image of the sounding of seven trumpets, which heralded the rise to prominence of the religion of a ten-horned Beast (a new and more terrible form of the Dragon), which had arisen out of the Sea. By false miracles the two-horned Beast would seduce people into worshipping the ten-horned Beast, thus instituting a new religion on earth; most Protestants understood this to refer unambiguously to the rise of Roman Catholicism.

Whereas most Protestant interpreters had understood a further image of the pouring of seven vials of wrath on the idolatrous followers of the Beast to refer to the history of the Protestant Reformation, Newton 'synchronised' each 'correspondent' vial and trumpet, and in turn harmonized these with the image of seven thunders. He argued that the last was added so that the 'intervals' between the seven vials, trumpets, and thunders might depict the same mystery (666) as the name of the number of the Beast. Thus each numerically linked vial and trumpet offered two different accounts of a particular period, each image enriching the picture offered by the other. By not reserving a special place for the vials as a specific account of the trials of Protestantism, Newton clearly implied that the Reformation had hardly made a dent in the growing power of the bestial empire.

At the sounding of the fifth trumpet the power of the Beast grew dramatically, and he made war upon the 'remnant' of the Woman's seed. For most Protestant exegetes – and Newton was no exception – this prophetic moment heralded a lengthy period that was depicted by an array of the most vivid images to be found in the Apocalypse. This was the period of the reign of the Man of Sin, or Antichrist, described in Revelation as the False Prophet or two-horned Beast, the last of which would morph into the Whore of Babylon. As Newton explained of the two-horned Beast: 'His being a heathenizing christian Ecclesiastical State makes him ipso facto a Whore in the strictest sense, & we have no reason to suppose more Apocalyptic Whores than one.' This period, up to the end of the sixth seal, lasted (in Revelation) for 1,260 days, during which the Woman, now fully in the wilderness, is kept in her place by the

9. ***The Whore of Babylon*****, according to Albrecht Dürer, 1498**

Beast. The latter makes war on and slaughters saints and martyrs, while the kings of the world fornicate with and worship the Whore.

According to Newton, the sixth trumpet (and for him, the sixth vial) referred to a period, the Great Tribulation, when the apostasy reaches its peak. The gospel is preached to every nation and the surviving remnant of the godly give thanks to God. The last trumpet and vial describe the arrival of many people from different nations bearing palms; the Lamb of God feeds them and sends them to living waters, while God wipes tears from their eyes. The Lamb is reunited with his wife to be, an image conventionally understood to be the reunion of Christ with the saints and martyrs.

Prophecy as history

Newton and his radical Protestant contemporaries were steeped in these and other prophetic images and for such individuals they made sense in their own right. However, for their full explication, they still needed to be 'applied' to historical events. Newton followed his definitions with an analysis of the history of the church that was alternatively expressed in the form of 'propositions' or 'positions', in a form reminiscent of a mathematical treatise. The fifth seal, for example, referred to the period when the Emperor Diocletian persecuted and slaughtered Christians at the start of the 4th century CE. The advent of Emperor Constantine ushered in the following seal, a period when Christianity became the state religion by dint (Newton believed) of diluting it to appeal to pagans. On Constantine's death in 337 the empire was split into East and West, the appearance of the latter (according to Newton) being the rising of the ten-horned Beast from the Sea.

Constantine's sons, who became leaders of these domains, differed in their religious views; one, Constans, was pro-trinitarian or as Newton termed it 'Homoüsian', while his brother (Constantius II) supported the Arian position, named after the priest Arius who had defended the lesser status of Christ with respect to God. By 364 the religion of the Beast was openly worshipped in the form of idols such as 'dead men's bones & other reliques of martyrs' and along with the worship of ghosts this soon became universal, Newton noted, 'as it hath continued ever since'. Now the devil was let loose on Earth to play what Newton called 'his cunning game', seducing ignorant people by means of false or diabolical miracles. In Newton's understanding of events, this was represented by the triumph of trinitarian Roman Catholicism and the persecution of godly Arians.

The Great Apostasy, accomplished by making Athanasian trinitarianism the official religion across the Roman Empire in 380, was described by the opening of the seventh seal. For Newton, the apostates, who were to overrun the visible church and persecute the godly, were to be Christians, albeit of a 'heathenish' and perverted sort; some might quibble with the idea that they were outwardly of the Christian faith, he argued, but a Christian 'was capable of being wors then any other sort of men'. The sounding of the first trumpet in 395 was synchronous both with an image depicting a terrible wind from the east and with the image of the first vial. This told of 'a noisome and grievous sore', which fell upon 'the men which had the mark of the beast, and upon them which worshipped his image'. Unwittingly, the writers of the early Catholic Church provided Newton with first-hand evidence of the great depravity of the clergy in this period, which led God to deploy hordes of Goths against them from the eastern part of the Empire. While Catholics bloodily persecuted groups of Christians who wanted to separate from the main Catholic Church, a practice Newton found the most deplorable of all, the Goths turned their attention to Rome itself in savage events, culminating in the sack of the city in 410, that were depicted by the second trumpet and vial.

Newton went on to claim that the third trumpet and vial, coincident with an image of a southern wind, depicted incidents in which African Catholics were slaughtered at the hands of Vandals. These were much more vicious than the Goths, who despite the occasional act of barbarity, had run Rome in a godly manner. From Victor's *History of the Vandalic Persecutions*, Newton learnt of the terrible atrocities meted out by the Vandals on the persecuting African Catholics. The latter were unprecedentedly bloody, Newton argued, and they murdered those who refused to follow their own superstitious practices, setting on foot 'those bloody persecutions which have been exercised in Europe & continue in the Roman Catholick Church to this day'. Again and again, Newton recorded in a state of great emotion, the Vandals paid back Catholic persecutions with interest. The Vandalic leader Genseric tortured nuns with red-hot iron plates, causing many deformities, and Newton agreed that this was 'very severe'. Nevertheless, the Catholics were unchaste, and it was divine justice that so many thousands suffered. Crucially, he argued, the Vandals persecuted them for their immorality and not for their religion.

The worship of images and of the Virgin 'came in' at the end of the fourth trumpet and vial, and God now briefly permitted the African Catholic Church to be restored, so that its insolent and stubborn clergy could be persecuted again and again by the Vandals. At the start of the fifth trumpet and vial, there was 'a new scene of things'. Revelation described how smoke arose from a pit from which swarmed a plague of locusts – armies in prophetic language – who were to torment no living thing except those who did not have the seal of God on their foreheads. This was the rise of Islam, to be dated from when Mohammed found his vocation as a prophet in 609 CE (as Newton dated it). His flight from Mecca to Medina in 622 was the opening of the pit, but the fifth trumpet and vial properly lasted from 635 to 936. The extended torment referred to the fact that Muslims had repeatedly laid siege to Constantinople without being able to take it. Elsewhere, transubstantiation and the canonization of saints 'came in' to Catholicism, as the apostasy reached its highest point.

In the sixth trumpet and vial, an angel lets loose four others bound up in the River Euphrates to prepare for the slaying of the 'third part of mankind' by horsemen with bright breastplates sitting on horses with heads like lions. Those who continued to worship ghosts and idols of gold, silver, brass, stone, and wood 'which neither can see nor hear nor walk' were condemned, along with those who did not repent of their sorceries, fornications, and thefts.

The sixth vial told of how the Euphrates dried up while from the mouths of the Dragon, the Beast and the False Prophet came three unclean spirits in the form of frogs, the spirit of miracle-working devils who prepared kings and gentiles for the battle of Armageddon. To Newton, and many of his Protestant contemporaries, the trumpet depicted the rise to power of the Turkish Empire, along with Saracens 'the very great scourges of the Christian world for this last thousand years'. The capture of Constantinople in 1453 was the slaying of the third part of mankind, while no one needed to ask who was described by the prophecy's indictment of false worship.

This astonishing and utterly original analysis was the overwhelmingly important concern of Newton in the 1670s and 1680s. Using similar techniques to his radical Protestant contemporaries he totally inverted what orthodox Christians of all persuasions took to be the heroes and villains of history. Indeed, at exactly the same time that he wrote his *Principia*, Newton was composing a detailed and extensive analysis of the way in which Catholics – whom he termed 'sorcerers' and 'magicians' – fulfilled the conditions of the sixth trumpet and vial. Many of the events that were to precede the final trumpet and vial were yet to come, and Newton repeated the caution of his philosophical work by stating that he would not hazard shaky conjectures about the exact nature or timing of future events. Rather, he saw his work as an observational and evidence-based analysis of how prophecies had been fulfilled throughout history. In the early 18th century he pushed back even further the great events of the future, believing that a long period of corruption had to take place before the Second Coming.

Chapter 7
The divine book

In early June 1679 Newton's mother died from a fever apparently caught while tending his half-brother Benjamin. Having dealt with his mother's illness and the business of the estate for about six months – not forgetting the many hours he spent daily on theological matters – he returned to Cambridge at the end of November. The day after he returned from Woolsthorpe, he replied to a letter from Robert Hooke. Innovative scientific entertainment had virtually ceased at the meetings of the Royal Society and, as secretary, Hooke implored Newton to communicate anything 'philosophicall' that might occur to him. Momentously for the development of Newton's orbital dynamics, he asked Newton what he thought of his theory of analysing planetary motions by means of an inertial path coupled with a force directing one body to the centre of an attracting body.

In reply Newton pleaded that he had given little thought to philosophy for many years, 'out of applying myself to other things', but offered a small 'fansy' concerning the Earth's daily motion. If an object fell to Earth, its diurnal rotation would not cause the object to fall behind the point directly beneath it ('contrary to the opinion of the vulgar'), but its west to east motion being greater at the height from which it was dropped than at positions closer to the Earth, it would fall *in front of* its original position (the east side). If an object were dropped from a tall tower,

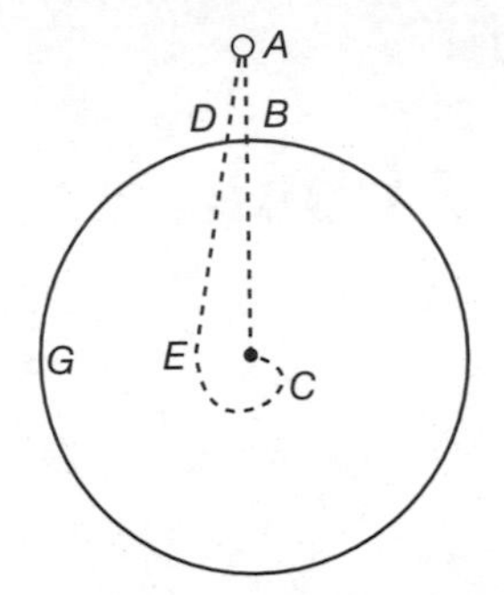

10. Newton's suggested path for an object dropped vertically from above the Earth's surface. The object's path is assumed to continue inside the Earth's surface as the Earth revolves around C anticlockwise (i.e. BDG).

diurnal rotation might thereby be proved and on the assumption that the Earth offered no resistance he drew a diagram detailing the spiral path of the object towards its centre.

Hooke responded that, instead of a spiral, on his supposition of inertial motion plus centrally directed attraction, a body such as Newton described would carve out an elliptical figure. This would forever move according to the curve AFG except where it encountered resistance and fell closer to the centre of the Earth. This perceptive comment, aired in one of Hooke's earlier publications, has justly caused historians to feel that Hooke deserved far more credit than Newton and subsequent commentators have granted him in forging the basic elements of orbital dynamics; however, it remains true that he could never demonstrate how the elliptical motion of orbiting bodies resulted from his physical principles.

Unable as ever to be corrected, Newton replied that, again assuming no resistance, the figure would not be an ellipse but that the object would 'circulate with an alternate ascent & descent made by its *vis centrifuga* & gravity alternately overballancing one another'. Newton's answer shows how far

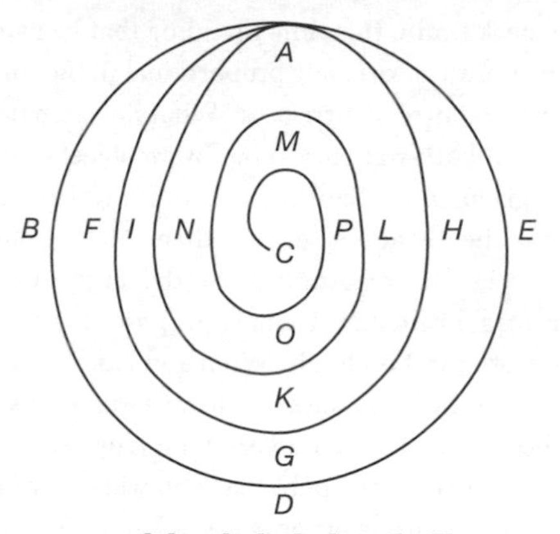

11. Hooke in turn argued that the body described by Newton would revolve in the ellipsoid AFGHA, unless it experienced some resistance, in which case it would descend close to the centre of the Earth

away he was from the analysis of celestial motions he would adopt seven years later in the *Principia*, but he also hinted at a much more sophisticated way of dealing with the problem according to continuous and infinitesimally small elements of gravitational force. Moreover, he implied that he could deal with a force of gravity that did not remain constant but varied from the centre outwards.

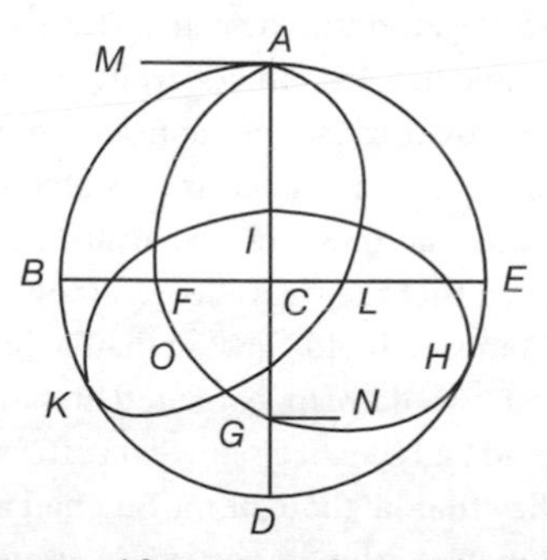

12. Newton's response, with gravity and 'centrifugal force' alternately overpowering each other

Hooke wrote back again, this time revealing that he had supposed that gravity was always inversely proportional to the square of the distance from the centre of attraction. What now remained, he said, was to show what path was carved out by an object centrally attracted by a given body according to a force inversely proportional to the square of the distances between them. Having offered Newton the crucial hint regarding a new dynamics of rectilinear inertia and central attraction, Hooke now posed a pertinent question (discussed in London by Wren and Hooke over a number of years) regarding how to relate the inverse square law to a planetary orbit – known from Kepler's First Law to be an ellipse. He told Newton that he had no doubt that 'you will easily find out what that Curve must be, and its proprietys [sic], and suggest a physicall Reason of this proportion'. Despite his later dismissal of Hooke's abilities, and his refusal to continue the correspondence any further, Newton later confessed to Edmond Halley that this exchange had incited him to think anew about celestial mechanics. Indeed, it was probably about this time that Newton momentously used Kepler's Second Law to demonstrate that on an elliptical orbit a body is subjected to an inverse-square law of attraction.

Another correspondence, this time with the first Astronomer Royal, John Flamsteed, was equally significant in the development of Newton's thinking about celestial motions. Early in November 1680 a brilliant and, to many, frightening comet (the so-called Great Comet) became visible to astronomers, while another appeared the following month. Partly because they appeared so infrequently, the status and orbits of comets was unclear to contemporaries. Descartes had argued that they were exhausted suns, while most astronomers believed that they travelled in straight lines. However, on 15 December Flamsteed told Newton that he had predicted that the November comet would reappear and that, having looked for it a few days earlier, he had seen it again. Soon afterwards Flamsteed told Edmond Halley that he thought the Sun had attracted the comet – a dead planet – within its vortex. He argued that the comet was turned *in front of the Sun* from its original southwards path by the attraction of the north pole of the Sun, but it was also moved sideways by the rotation of the solar vortex (from e to g in Figure 13). The Sun continued to attract the comet to its centre but at the same time the anticlockwise vortex constantly changed the path of the comet. When it came closest to the Sun (at C), the comet was sufficiently twisted by the vortex that it presented its opposite 'side' to the Sun, and the attractive force was turned into a repulsive force. The tail, he argued, was caused by the sun heating the humid parts of the atmosphere.

Fascinated by the comet, Newton observed it from 12 December 1680 until it disappeared in early March 1681, deploying more powerful telescopes as the object faded. Unable to accept that the two comets were the same, he offered some incisive criticisms of Flamsteed's views at the end of February. Newton remarked that, although he could conceive of the Sun continuing to attract the comet to make it deviate from its original path, it would never attract the comet in such a way that it would end up being directly attracted in the direction of the Sun. Moreover, the solar vortex would only push the comet further away from the Sun. But even if a

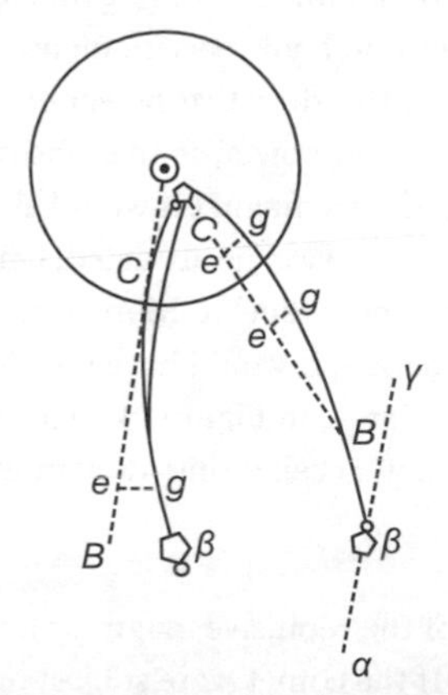

13. Flamsteed's suggested path for the comet of winter 1680–1. This begins at β on the lower right of the picture, and is repelled in front of the Sun at C

single comet had turned in front of the Sun, it would not have returned in the path that had been seen by astronomers. Moreover, on the assumption that the November and December comets were the same, another problem arose with the substantial length of time that had elapsed between the last sighting of its first appearance, and the first view of its second.

The only solution to these problems, Newton suggested, was to imagine that the comet had turned around *on the other side of the Sun* – but then the physical mechanism for this was unclear. He accepted that the Sun exerted some centrally attracting force that bent planets away from the straight line they would otherwise have taken, but this could not be magnetic since hot loadstones (natural magnets) lost their power. More importantly, even if the attractive power of the Sun were like a magnet, and the comet like a piece of iron, Flamsteed had still not explained how the Sun would suddenly switch from attracting to repelling.

The magnetic account had offered the best explanation of the Sun's power over planets for nearly a century. Newton's complete rejection of it, based on an understanding of magnets that went back to his 'Questions' notebook, was momentous. In a further letter he remarked that the 'directive' power of a magnet was stronger than its 'attractive' power, so that once an object was in a position to be attracted by a magnet it would always remain in that position and would thus always be attracted. Once it attracted the comet, the Sun would never repel it. Moreover, even if a repulsive magnetic force did operate, it would have repelled the comet some time *before* perihelion (at K, in Figure 14). The comet would have continued on its journey, accelerating away from the Sun on its other side.

Newton's dismissal of the repulsive magnetic force was, as usual, immensely original. If the comet were subject only to a continuous attractive force, this would decelerate the comet as it left the Sun and make the comet travel along an orbit close to that observed.

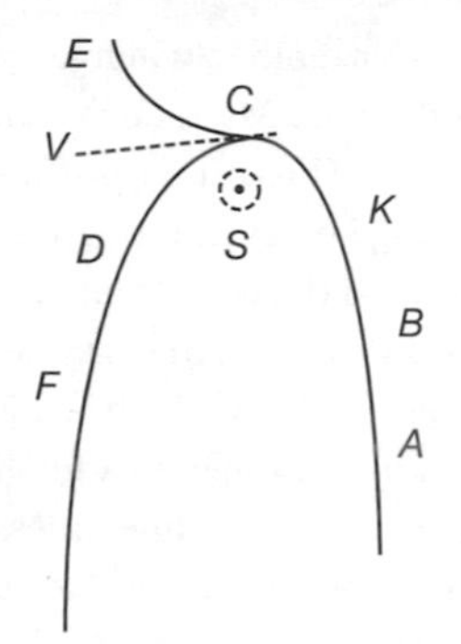

14. Although he still believed that the comets of November and December 1680 were different, Newton's crude but ingenious diagram displays a possible path for a single comet behind the Sun

Perhaps it was only at this point that Newton saw that a one-comet solution, using only an attractive force was viable. However, in the letter to Flamsteed, using the same term as he had mentioned in his letter to Hooke, Newton argued that *vis centrifuga* or 'centrifugal force' had 'overpower'd' the attraction at perihelion, allowing the comet to recede from the Sun despite the attraction. Although Newton would dispense with this notion of centrifugal force as the tendency (or measure) of an orbiting object to move away from the attracting body, the notion of continuous attraction would be a cornerstone of the more mature dynamics of the *Principia*. He was close to – but still three years away from – realizing how comets could be treated like any other heavenly body.

The motion of orbiting bodies

When Edmond Halley visited Cambridge to see Newton in August 1684, it was the result of discussions about celestial dynamics that had been taking place amongst the virtuosi in London for some time. According to Newton, when Halley asked him what curve would result from an inverse-square force law, Newton immediately replied that he had calculated it to be an ellipse. However, when Newton searched for the demonstration he could not find it, and

Halley had to wait until November, when he received a short mathematical tract entitled *De Motu Corporum in Gyrum* ('On the motion of bodies in orbit'). The cosmos outlined in *De Motu* was an abstract system of moving bodies that obeyed certain mathematical laws. Newton now invented the term 'centripetal' to describe the centrally attracting forces working in his system, and defined as an 'innate force' that power by which a body 'endeavours to persist in its motion along a right line'. Linked to a further claim that bodies continue to infinity along a straight line unless otherwise acted upon, this would be the basis of the first Law of Motion in the *Principia*. Under the heading 'Hypothesis 3' he also described an early version of the 'parallelogram of forces' rule that ultimately became the second Law of Motion in the *Principia*.

Central to his analysis was his demonstration in 'theorem 1' of *De Motu*, of Kepler's Second Law, by which objects swept out equal areas in equal times, applied to all bodies orbiting about a centre of force. This analysis divided up the area carved out by the orbiting motion into infinitesimally small elements, the orbiting body being subject at each moment to 'impulses' that changed the direction of the body an infinitesimally small amount and created a series of infinitesimally small triangles, each with the same area. However, theorems 2 and 3 dealt not with impulses but with continuous forces, which ultimately could be treated in terms of the formula for continuous (uniform) acceleration discovered by Galileo. The tension between these two different accounts of force, an 'impulsive' one measured by mass times velocity (mv = momentum) and the other, a 'continuous' version measured by mass times acceleration (ma), would remain in his *Principia*.

Theorem 3 showed that orbiting bodies were subject to an inverse-square force, and Newton went on to demonstrate that planets were such bodies, revolving around the Sun according to the laws outlined in his tract. Momentously, under 'Problem 3' he proved that an inverse-square law governed the path of bodies that moved in elliptical orbits. Furthermore, for the first time comets

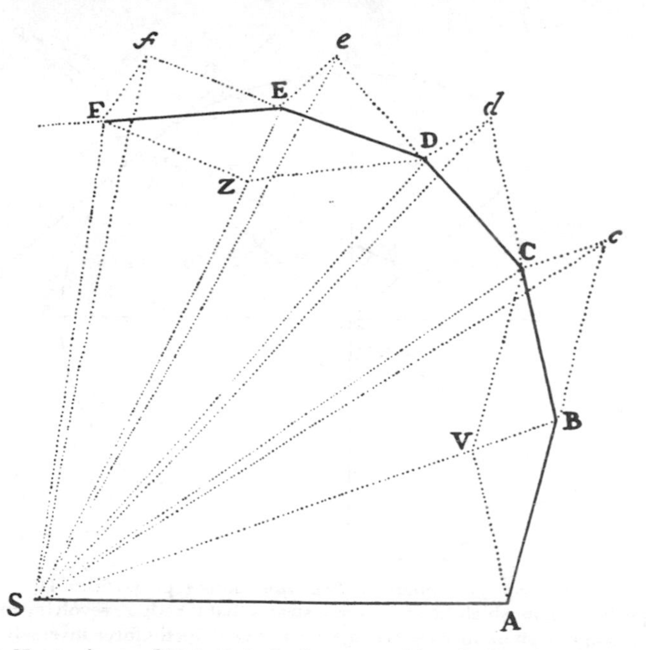

15. Newton's proof (*Principia*, book 1, proposition 1) of Kepler's Second Law. A body orbiting along the path ABCDEF, attracted by a centripetal force in the direction of S, can be thought of as being subjected at equal moments of time to a 'single but great impulse' successively at B, C, D, etc. The distances between these points can become indefinitely small so that the orbit becomes a curve. Since SAB, SBC, etc., are equal triangles, the body will sweep out equal areas in equal times.

were incorporated into a universal system of mathematical natural philosophy and he argued that it was even possible by close analysis to determine whether they were periodic (i.e. had elliptical orbits and thus returned at regular intervals). Under 'Hypothesis 1' he noted that the bodies in his system moved through non-resisting media, although he did add some material on motion in a resisting medium in the form of 'Problems' 6 and 7.

A fascinating correspondence with Flamsteed over the winter of 1684–5 shows that Newton was already trying to link his analysis to

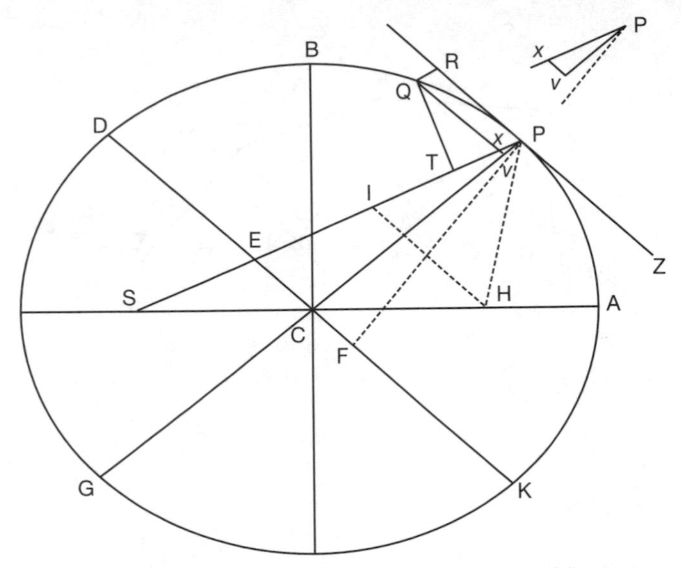

16. The diagram accompanying *Principia*, book 1, proposition 11, problem 6, in which Newton demonstrates that a body P, revolving in an ellipse around focus S, is subjected to a centripetal force inversely as the square of the distance SP

a more precise view of the actual motions of the planets and their satellites, as well as of comets, and that he was testing the accuracy of Kepler's Third Law. Flamsteed, who had read *De Motu*, was aware that Newton's November tract implied that planets could be treated, like the Sun, as centrally attracting objects. In the mean time Newton had gone further, assuming that if Jupiter governed the motions of its satellites then it also had an effect on other planets – and vice versa. He asked for data concerning Jupiter's 'action' on Saturn in a letter of December 1684 but Flamsteed – still thinking that any such force would have to be magnetic – baulked at the idea that planets could influence each other over such long distances.

When Newton composed a revision of *De Motu* in early 1685, the 'Hypotheses' had been elevated to the status of 'Laws'. Although he was still some way away from his theory of Universal Gravitation, he now made the revolutionary claim that because of the vagaries of numerous mutual planetary interactions, the centre of gravity of the solar system was not always in the same position as the Sun, and the orbits of planets were thus always irregular, or never exact Keplerian ellipses. Planetary orbits, which had for centuries stood as exemplars of unchanging perfection, were, in fact, constantly undergoing minute changes. The human intellect, Newton remarked, was incapable of dealing with the complexity of real motions but for the most part planetary orbits could be treated as elliptical. Later he would argue that the stability of such a system could only be due to the hand of a divine geometer. At this point he also introduced an argument that would be crucial for his approach in the *Principia*, namely, that since comets experienced no visible diminution in their tails, there was nothing actually existing in the free spaces of the cosmos to resist their paths. Newton now also began to consider whether a very fine aether that offered no resistance could in any sense be said to exist *at all*.

A dramatic change in his analysis of the way a force alters a moving body now allowed him to reintroduce a generalized nation of inertia, namely that a body remains in its current state of motion or rest, *both of which were relative to whichever system was chosen as a frame of reference*. Major and revolutionary insights followed. Armed with the relativized concept of inertia, Newton announced in a further set of 'definitions' (written after the revision to *De Motu*) that uniform circular motion around a centrally attracting source was not an example of simple inertial motion but was in fact the compound of the object's velocity and a continuously attracting force that caused it to deviate from the path it otherwise would have taken.

The relativistic implications of the notion of inertia raised the thorny question of whether absolute motion could be detected, a problem that harked back to the analysis in 'De Gravitatione'. Realizing both the theological and scientific implications of the problem, Newton argued forcefully in an addition to the definitions for the existence of an absolute space independent of the things within it, 'since all phenomena depend on absolute quantities'. As we have seen in his remarks to Burnet, he believed that ordinary people experienced the world in relative terms, and it was right that prophets should speak to them in that language. In the addition to the revision of *De Motu*, Newton remarked that 'ordinary people who fail to abstract thought from sensible appearances always speak of relative quantities so much so that it would be absurd for wise men or even Prophets to speak to them otherwise'. Without the reference to theology, this significant view made its way into the *Principia*, where the vulgar were said to consider quantities only as they related to 'perceptible objects'. However, Newton went on, 'in philosophical discussions, we ought to step back from our senses, and consider things themselves, distinct from what are only perceptible measures of them'. Nevertheless, Newton's efforts to show that one could detect an 'absolute' frame of reference that was privileged above any others would ultimately turn out to be illusory.

In the same draft Newton added six 'laws of motion', the third of which announced that 'as much as any body acts on another, so much does it experience in reaction'. This effectively set up an equality between the force by which a body 'resisted' being moved (later the 'vis insita' described in Definition 3 of the *Principia*) and the 'impressed' force that was imparted to any body, whether continuously or by impulses. It was an early version of the third Law of Motion in the *Principia*, and, along with his notion of mass, it now gave him the tools that would allow him to generalize the notion of centripetal attraction to all bodies in the universe.

Newton now defined much more precisely the quantity of 'bulk' of matter, which he at first asserted was 'usually identical' to its gravity. In a revision of the 'definitions' written in the spring or summer of 1685, he defined the 'quantity of matter' (or 'mass'), as basic and undifferentiated matter, so that a body 'twice as dense in double the space' would have four times the quantity of mass. Perhaps most significantly, the new analysis demanded that all basic matter was essentially identical, and where there was no matter there was nothing at all. In book 3 proposition 6 of the final version of the *Principia* (1687), he implicitly linked the mathematical concept of 'mass' to his alchemical analysis by introducing a 'Hypothesis 3' in which he asserted that, because the fundamental building-blocks of matter were all the same, all forms of matter could in principle be 'transmuted' into each other.

The ancient Newtonians

By November 1685 he had completed a draft of the *Principia*, also entitled *De Motu Corporum*. This consisted of two books, the so-called 'Lectiones de Motu' (or 'Lectures on Motion') and 'De Mundi Systemate' ('On the system of the world'). The 'Lectiones' expanded the demonstrations in the initial forms of *De Motu* (and their revisions), and Newton attempted to deal with the intractable problems involved in considering the mutual attractions existing between more than two bodies.

Over the winter of 1685–6 he expanded the 'Lectiones' by enhancing his analysis of the motions of a satellite (an abstract body but bearing properties virtually identical to the Moon) under the influence of two or more bodies (again abstract, but clearly meant to be the Sun and the Earth), and added a masterful Proposition (XXXIX) that referred unambiguously to his knowledge of the calculus. Undoubtedly this was in some degree to assert his independence in the development of the calculus from the work of Leibniz, who had published its basic axioms for the first time in 1684. In early 1686 Newton expanded his treatment of motions in resisting media, an analysis that became so large that he split it off to form a text that would become book 2 of the *Principia*. The first part, an analysis of motions in non-resisting media, would become book 1. In the final version of book 2, Newton added more complex material on pressure and viscosity, arguing that the existence of Cartesian-type vortices was physically impossible.

Newton began the second book of the 1685 work, 'De Mundi', with a reference to the philosophy and astronomy that underpinned the work of Plato, Pythagoras, and the Roman king Numa Pompilius. As a symbol of the figure of the world with the Sun in the centre, Numa had 'erected a round temple in honor of *Vesta*, and ordained a perpetual fire to be kept in the middle of it'. In arguing that there had been knowledge of the natural world that had become lost, Newton was following the majority of his contemporaries. In a large treatise from the mid-1680s ('The Philosophical Origins of the Gentile Theology'), he argued that the Ancients had believed in a heliocentric cosmos but that this had been perverted by misinterpretation. Whereas Pythagoras and others had correctly understood the true meaning of symbolic representations of a heliocentric cosmos, with a central Sun encircled by the concentric orbits of the planets, Greeks such as Aristotle assumed that the central object in such a scheme was the Earth.

Via Orpheus and Pythagoras, the Greeks had originally received their understanding of the natural world from the Ethiopians and Egyptians, who had concealed these truths from the vulgar. At this time there was a 'sacred' philosophy – communicated only to the cognoscenti – and a 'vulgar' version, promulgated openly to the common people. The Egyptians

> designated the planets in the [correct] order by means of musical tones; and to mock the vulgar, Pythagoras measured their distances from one another and the distance of the Earth from them in the same way by means of harmonic proportions in tones and semitones, and more playfully, by the music of the spheres.

In 'De Mundi' Newton repeated the view that, like the Chaldeans, the Egyptians had known that comets were heavenly phenomena and could be treated as if they were a sort of planet.

The Egyptians built temples in the form of the solar system and derived the names of their gods from the order of the planets. As such, the ancient religion was modelled on the understanding of the heavens and Newton occasionally referred to the Ancients' 'astronomical theology'. If one added the seven known planets (including the Moon) to the five elements – air, water, earth, fire, and the heavenly quintessence – then one arrived at the twelve basic gods that were common to all the ancient religions. Noah was Saturn and Janus, and had three sons. Newton followed other historians in adopting a 'euhemerist' approach by means of which pagan myths were held to refer to real people who had been deified by different nations under different names. Evidence for this included a similarity of names, and in particular the fact that the descriptions of their characters and deeds were ostensibly identical.

The first part of 'De Mundi' was thus an offshoot of a much larger project that was well under way by the time it was written, and Newton's vast effort to identify 'hints' of the true philosophy in ancient writings evolved in many different ways over the following decades. In a slightly later work entitled 'The Original of Religions', for example, he asserted that the ancient Chinese, Danes, Indians, Latins, Hebrews, Greeks, and Egyptians all worshipped according to the same practices, while Stonehenge in England was clearly another vestal temple. Nothing could be more 'rational', Newton added, than this aspect of religion: there was no way 'without revelation to come to the knowledge of a Deity but by the frame of nature'. Armed with knowledge of the true philosophy, he could recover the sacred philosophy that had been 'veiled', while this in turn would guarantee that his own account of the world was true. Just as his theological work aimed at the restoration of the true religion, so he always believed that his scientific work was essentially an effort to restore a lost knowledge.

Principia

While the 'Lectiones' dealt with an abstract mathematical system, the rest of 'De Mundi' dealt with data on tides (gleaned from Flamsteed in correspondence of autumn 1685), pendulum experiments, the real Moon, and other phenomena from the real world. It was by comparing these with the mathematical world described in the 'Lectiones' that Newton could assert that the laws that operated in the abstract world also governed phenomena in our own. However, at this point he lacked an adequate account of comets for the second book and he worked vigorously over the winter of 1685–6 to produce one.

The final part of the *Principia*, book 3, was completed early in 1687 and dealt with the actual system of the world. From basic principles, and from astronomical and physical data, Newton demonstrated that the earth was flattened at the poles (i.e. was an oblate spheroid). Finally, he showed how his physics could explain the action of comets, whose orbits could be treated as parabolas close to the Sun and ascertained by exact measurements, although the search for part appearances of comets with similar profiles might show that they were periodic and thus elliptical. Newton included some fascinating passages on the function of comets. As these approached the Sun their tails were replenished by solar material, which in turn rejuvenated the fluids that provided sustenance to living things whenever the planet passed through the tails. Newton added that the purest part of the air, which sustained all life on Earth, also came from comets. Evidently this was an extended version of the terrestrial circulatory system described in his alchemico-philosophical work of the 1670s.

When it appeared in 1687, the *Principia* boldly announced a credo that would influence the practice of science for the next three centuries. Hypotheses were to be banished, and well-designed experiments were to be made the basis of general mathematical laws. These laws were to be as few in number as possible and were to be assumed true everywhere, unless counter-evidence could be found. Its crowning conceptual glory was the law of Universal Gravitation, which held that massive bodies attracted each other according to a constant 'G', multiplied by the product of the masses and divided by the square of the distance between them (Gmm'/r^2). The epoch-making implications of this work now became clearer: massive planetary bodies could no longer be privileged as the sole bearers of centripetal attractions, since from the third law of motion all massive bodies exerted such a force. The stunning conclusion was that each and every massive body in the universe attracted every other body. This was to raise substantial problems for Newton and his contemporaries. What was attraction? How, for example, could it be exerted from one end of the universe to the other? Through what sort of medium did it operate?

Shortly before dispatching the final parts of the book to London, Newton composed a remarkable 'Conclusio' to the work that promised to extend his analysis of the *Principia* to all other terrestrial phenomena. Based on the way he had used Universal Gravitation to explain macro-phenomena, he argued that short-range forces should be adduced to account for the 'innumerable' other local motions that could not be detected on the grounds of their size but that underlay a wide range of earthly phenomena such as electricity, magnetism, heat, fermentation, chemical transmutations, and the growth of animals.

As he had done for celestial motions in book 2, so now for terrestrial phenomena Newton dispensed entirely with the aether that in various guises had served him so loyally over the previous two decades. In its place, he proposed simply to use what he called attractions and repulsions and he invoked popular speech to say that the term 'attraction' was conventionally used to describe any force by which particles 'rush towards one another'. At short range, these forces were attractive and accounted for the 'condensing' properties he had noted in his previous alchemical and philosophical work. Further apart, forces were repulsive, accounting for the phenomena of surface tension (such as flies walking on water) explained by means of an aether in the 'Hypothesis'. However, Newton's claim that there might be a number of such forces put a strain on his demand that philosophers should adopt a minimum of general principles.

Newton added that he had mentioned these forces only as an incitement to do further experiments, but then offered a speculation that was based on his view that the basic stuff of matter was the same. Because most of space was empty, the forces that allowed bodies to cohere would make them coalesce into regular structures 'almost like those made by art, as in the formation of snow and salts'. Internally, there would be net-like structures formed by very long and elastic geometrical rods, a fact that explained how some bodies could be more easily heated or allow more light to pass through them than others. Again invoking quasi-alchemical concepts, he argued that, with attractive forces, different net-like arrangements of the basic elements of matter allowed transmutations to take place. Using Helmontian concepts he asserted that, by fermentation, 'that rare substance' water could be made or 'condensed' into the 'more dense substances' of animals, vegetables, and minerals, and finally 'into mineral and metallic substances'. On the other hand, repulsive forces gave rise to vapours, exhalations, and air if they were dense bodies and, if rarer, to light itself. Newton drew back from publishing his stunning conception of the microworld, and condensed it into a draft preface; this too failed to make the final version.

Robert Hooke – the great pretender

In May 1686, just after book 1 had been presented to the Royal Society, Halley told Newton that Hooke had 'some pretensions' to the inverse-square law and had claimed that he had brought this to Newton's attention. Although Hooke did not claim any rights to the demonstration that conic sections were generated from such a law, Newton's patience had run out for the last time. He told Halley that Hooke had pestered him throughout their correspondence of 1679–80 and had given him nothing he did not already know. A few days later, having pored over old papers, he angrily noted that

Hooke had only 'guessed' that the inverse-square law extended down to the centre of the Earth but in doing so had been in error; now, Newton told Halley, he had decided to suppress the third book. Philosophy was 'such an impertinently litigious Lady that a man had as good be engaged in Law suits as have to do with her'.

Newton did not stop here, telling Halley that upon finishing the main bulk of the letter he had heard that Hooke was making a 'great stir pretending I had all from him & desiring they would see he had justice done him'. As before, he pointed out where Hooke had stolen other people's work and passed it off as his own, writing in such a way

> as if he knew & had sufficiently hinted all but what remained to be determined by the drudgery of calculations & observations, excusing himself from that labour by reason of his other business: whereas he should rather have excused himself by reason of his inability.

Newton sarcastically noted that, according to Hooke,

> Mathematicians that find out, settle & do all the business must content themselves with being nothing but dry calculators & drudges & another that does nothing but pretend & grasp at all things must carry away all the invention as well of those that were to follow him as of those that went before.

In tenor with previous exchanges between the two, Newton wove a complex and wildly implausible tale of how Hooke might have gleaned the inverse-square law from his previous correspondence. In response Halley put Newton's mind at rest and told him that having discussed the matter in a coffeehouse, few others believed that Hooke had either the demonstration relating the elliptical orbit to the inverse-square law or a gigantic system of nature.

Evidently, Newton did not suppress book 3 of the *Principia*, although he did make it more mathematical and less accessible.

It may well have been partly to teach Hooke a lesson, although the development of its contents in any case demanded a more forbidding treatment. As a whole, the *Principia* became a byword for impenetrability, with many accomplished mathematicians trying and failing to get very far beyond the first few propositions.

Chapter 8
In the city

Before he had finished the last book of the *Principia*, Newton found himself involved in a new crisis. Soon after he ascended the throne in early 1685, the Catholic King James II began to relax laws and practices aimed at restricting the ability of Catholics to hold office or attend university. In February 1687 the vice-chancellor of Cambridge University received an order requiring the university to admit Father Alban Francis to a degree of MA at Sidney Sussex College, and Newton acted quickly against the perceived threat to the Protestant integrity of his university. In April 1687 he was one of eight 'messengers' deputed by the university to appear before an Ecclesiastical Commission headed by Judge Jeffreys, a one-time undergraduate colleague of Newton but now infamous for having recently sentenced to death hundreds of supporters of the Protestant Duke of Monmouth. On 21 April Jeffreys harangued the Cambridge eight in his customary style but gave them an extension to prepare their defence further. On 12 May, Newton, Babington, and six others were told that their 'sly insinuations' had invoked the anger of the Commission and Jeffreys sent them packing with the injunction to sin no more lest a worse fate befall them.

At a meeting to prepare for the confrontation with Jeffreys in April, Newton had pushed strongly for an uncompromising stand on the admission of Father Francis, and in a short essay he argued that the situation was too important for the university to trust James's promise to safeguard the Protestant religion (as king of England James, despite being a Catholic, was also notionally the defender of the Anglican Church). Indeed James could not make any such promise, first, because it was forbidden by the terms of his own religion and, second, because he could not in any case legally use his dispensing power to remove laws guaranteeing the centrality of Protestantism in England. Englishmen would not give up laws governing liberty and property; with even less reason should they give up those guaranteeing religion.

In another essay Newton went on to examine the limits of the king's dispensing power, finding that he lacked the power to dispense with laws when there was no necessity for him to do so. In an analysis that marked him out as a 'Whig', in whose radical circles he moved when he became an MP in 1689, Newton downgraded the king's powers below those of the 'people' who alone had the power to decide whether dispensing with laws was necessary. In further documents, this time prepared for the final showdown with Jeffreys, he argued that the delegates' stand had been taken to defend their own religion; Catholics and Protestants could not live 'happily nor long together' in the same university, and if the fountains of Protestant education 'be once dryed up the streams hitherto diffused thence throughout the Nation must soon fall off'.

By 1687 Newton's active life as Lucasian Professor came to a halt. Having performed in front of what may occasionally have been a non-existent class for nigh on a decade, in 1684 he deposited a manuscript on algebra in the University Library to fulfil his professorial obligations. Published by William Whiston in 1707 under the title *Universal Arithmetick*, Newton's work praised the reliance of ancient mathematicians on geometry while lambasting the introduction of equations and arithmetical terms into geometry by modern analysts.

James II fled England at the end of 1688 and the arrival of William of Orange (in what was to be called the Glorious Revolution) gave

Newton an opportunity to show his allegiance to the new regime. Although he was described in the most glowing terms on the voting slips, it was still something of a surprise when in January 1689 he was elected as one of the two MPs for Cambridge University in the Convention Parliament. In early February he voted with the majority of MPs who determined that James had 'abdicated' from the throne in his retreat and in the following weeks he served on a committee that drew up the wording for a bill concerning the toleration of various kinds of dissenters. Newton naturally supported the toleration of various shades of Protestantism and believed that the state should allow worthy Protestants of any denomination (such as himself) to hold office. When the bill on this topic was passed into law on 17 May as the Toleration Act, dissenters could freely engage in public worship. However, the sacramental element of the Test Act had not been repealed and freedom of worship was refused to Catholics and anti-trinitarians.

Newton suffered a further setback in the summer of 1689 when his candidacy for the provostship at King's College was turned down, despite strong support from the new king William III. Nevertheless, he hardly lacked for admirers and disciples. A number of individuals vied to be the editor of the next edition of his great work, while others devoted themselves to mastering the work's incredibly abstruse contents. In turn Newton doled out patronage to his followers, such as the Savilian Chair of Geometry at Oxford that he helped obtain for David Gregory. On the Continent, the *Principia* was hailed by eminent natural philosophers such as Huygens and Leibniz, although both thought that Newton had neglected the entire purpose of natural philosophy by failing to offer a physical explanation for 'attraction'.

Having sounded the death knell for vortices in book 2 of the *Principia*, Newton struggled to explain gravity. In the first half of the 1690s he showed Fatio de Duillier and David Gregory many of the revisions and corrections he was making to the *Principia*. Some of these concerned the physical cause of gravity, and in a series of ‘classical’ scholia to propositions 4 to 9 of book 3 he showed that Universal Gravitation and other doctrines had been known to the Ancients and could be divined from a serious reading of the poems of Virgil, Ovid, and others. In these revisions Newton claimed that Universal Gravitation operated by means of ‘some active principle’ that allowed the transmission of force from one body to another:

> and therefore those Ancients who rightly understood the mystical philosophy taught that a certain infinite spirit pervades all space & contains & vivifies the universal world; and this spirit was their numen, according to the Poet cited by the Apostle: In him we live and move and have our being.

By the symbol of Pan and his pipes the Ancients referred to the way this spirit acted upon matter, ‘not in an irregular way, but harmonically or according to the harmonic ratios’. Much later, Catherine Conduitt noted that Newton thought that gravitation depended on mass, in the same way that sounds and notes depended on the size of strings.

This was not the only aspect of Newton’s general effort to restore the lost knowledge of the earliest times; at about the same time, he threw himself into a gigantic mathematical enterprise that purported to ‘restore’ the lost geometry of the Ancients. In late 1691 he also began to compose a text entitled ‘De Quadratura curvarum’, an extraordinary work in which he ranged back over his discovery of the calculus and his development of infinite series. Given that he drew heavily on his letters to Leibniz from the mid-1670s, it is clear that his principal aim was to assert his priority and superiority over him. When Gregory saw it in 1694 he remarked that Newton developed the theory of quadratures (integration) ‘astonishingly [and] beyond what can be readily believed’.

The years after the completion of the *Principia* witnessed some of the most intense intellectual activity of Newton’s life. In the late

1680s he planned to produce a work on optics in four books, intending in the concluding book to show how optical effects acted according to small-scale attractive and repulsive forces. In a draft, he repeated his remarks in the suppressed preface and conclusion of the *Principia* to the effect that philosophers should assume that similar kinds of force operated in the micro- as well as in the macro-world. However, he went on, this 'principle of nature being very remote from the conceptions of Philosophers I forbore to describe it in [the *Principia* lest it] should be accounted an extravagant freak'. Whatever Newton's original plans, he had reduced the proposed text to three books by 1694 and *Opticks* ultimately appeared in this form a decade later.

In the summer and autumn of 1690 he researched furiously into the vexed question of how Catholics and trinitarians had corrupted the true text of the New Testament. Due to a relaxation of the licensing laws governing publication in 1687, a number of anti-trinitarian works had appeared in print. When in 1689 the Catholic Richard Simon published a work that analysed the part of the central trinitarian text 1 John 5: 7–8 known as the Johannine comma, John Locke, a recent acquaintance of Newton, asked him for his views on the passage. In November 1689 Locke (who was about to publish his great works A *Letter on Toleration*, *An Essay Concerning Human Understanding*, and *Two Treatises on Government*) received Newton's lengthy exposition on both this and another trinitarian passage 1 Tim. 3: 16. There can be no doubt that he understood Locke to be sympathetic to his views, despite the thick veil of objective research with which Newton tried to conceal his work.

Simon had remarked that the passage's authenticity was guaranteed by Catholic tradition, even though it was not found in the oldest Greek manuscripts. Newton told Locke that it was yet another Catholic corruption, but that although they knew this, many humanists and Protestants had preferred to keep the text as it was a key piece of evidence against heretics. What he was about to do, he remarked disingenuously, was 'no article of faith, no point of discipline, nothing but a criticism concerning a text of scripture'. In short, the Church Father Jerome had inserted the false passage into his Vulgate and afterwards

> the Latines noted his variations in the margins of their books, & thence it began at length to creep into the text in transcribing, & that chiefly in the twelfth & following Centuries when disputing was revived by the Schoolmen.

After the advent of printing, it 'crept up out of the Latine into the printed Greek against the authority of all the greek MSS & ancient Versions'.

Newton's approach to these corruptions was threefold. First, he could show how and why the text was inserted into various manuscripts and printed texts. This involved a convoluted, scholarly analysis of texts in which he argued that trustworthy authors before Jerome would have referred to the text if it existed, but had not done so. There was no evidence that it was present in the oldest Greek texts and indeed some contemporaries had accused Jerome of inserting it according to his own whim. Newton himself put Jerome on trial and unsurprisingly found him guilty. Second, he actually had access to ancient manuscripts, and to printed editions that referred to manuscripts where the offending text was missing or flagged as problematic. If there were texts where the comma appeared, then Newton tried to show that they were written much later. Third, the restored, authentic passage apparently made more sense, and he recast the disputed text for Locke's benefit.

Soon afterwards, Newton sent Locke an account of many more problematic texts, 'for the attempts to corrupt the scriptures have been very many & amongst many attempts tis no wonder if some have succeeded'. According to Newton, all these corruptions had been initially made by Catholics '& then to justify & propagate them

[they] exclaimed against the Hereticks & old Interpreters, as if the ancient genuine readings & translations had been corrupted'. Scholars in this period lurched from one disgraceful act to another: 'such was the liberty of that age that learned men blushed not in translating Authors to correct them at their pleasure & confess openly that they did so as if it were a crime to translate them faithfully'. Protestants now collaborated in the crime, and Newton sanctimoniously told Locke that all these deceptions 'I mention out of the great hatred I have to pious frauds, & to shame Christians out of these practises'.

Breakdown

In 1692 and early 1693 Newton became extremely close to Fatio de Duillier, who pestered the older man with tales of the marvellous cures that could be effected by an alchemical potion developed by one of his friends. In one letter he asked Newton to invest a substantial amount of money in developing and marketing the product. In the early summer of 1693 Newton went from Trinity to London on a number of occasions, presumably to discuss this and other matters with him. By July Newton was in the throes of a breakdown, an experience that only became known when he sent a letter to Samuel Pepys in the middle of September. In this strange offering, composed while he was still in a great deal of turmoil, Newton was deeply concerned to deny that he had ever tried to use either Pepys or James II as a patron, and he told Pepys that he would have to withdraw from his acquaintance and indeed never contact any of his friends again. Locke received an even more troubling letter, written three days later from a pub in Shoreditch. Like Pepys, this was the first Locke had heard of Newton's concerns. Newton apologized for accusing Locke of trying to 'embroil' him with women, and begged forgiveness for wishing that Locke would die from a sickness from which he had been suffering. He was sorry for accusing Locke for being a Hobbist (i.e. a materialist) and for saying that Locke undermined the basis of morality in his *Essay*.

Despite these egregious insults, Pepys and Locke reacted with admirable understanding, and indeed Newton soon claimed that he had forgotten what he had written. Overwork, mercury poisoning, repressed attraction for Fatio, and a failure to get a job in London have all been offered as explanations for Newton's bizarre behaviour but no single explanation seems to be convincing.

As he recovered his equanimity and normal life resumed, Newton had one last try to rectify some of the problems that had dogged his treatment of lunar theory in the *Principia*. Arguably, it would be his last major sustained scientific undertaking. From the summer of 1694 he attacked the issue again, and he visited John Flamsteed at Greenwich to acquire the latest data. Flamsteed agreed to let Newton see his upgraded lunar observations, but added the rider that Newton had to promise not to show them to anyone else. In turn, Flamsteed wanted the corrections to his observations that Newton claimed he could make in virtue of his improved theory. Nevertheless, Newton was not about to treat Flamsteed as an equal, virtually demanding that the Astronomer Royal send him his raw observations according to his bidding. As it turned out, the three-body problem Newton had to solve in order to make headway with the problem proved too difficult for him, while Flamsteed struggled to provide observations of the type and precision that Newton demanded.

In an atmosphere of increasing mutual suspicion, Flamsteed heard that Newton was showing his own 'corrections' of the data to Halley and Gregory, while Newton took umbrage at Flamsteed's alleged sloth in providing raw data and also at his wish to know the theoretical basis of Newton's emendations. Over the following years the relationship deteriorated still further. When Flamsteed threatened in 1698 to reveal in print that he was providing the data with which Newton could improve his theory, the latter exploded with rage and prevented publication. Immersed in his role as Warden of the Mint and unwilling to have a wider audience reminded of his failure to solve the Moon's motions, he told

Flamsteed that he did not care to be 'publickly brought upon the stage about what perhaps will never be fitted for the publick & thereby the world put into an expectation of what perhaps they are never like to have'. He did 'not love to be printed on every occasion' and much less 'to be dunned & teezed by forreigners about Mathematical things or to be thought by our own people to be trifling away my time about them when I should be about the Kings business'. Their relationship, always shaky, could never recover.

Chapter 9

Lord and master of all

Newton's efforts to find a suitable post in the metropolis finally bore fruit in 1696, completing his bizarre transition from hermit to senior civil servant. His erstwhile Trinity colleague Charles Montagu (Baron Halifax after 1700) signed a letter confirming Newton's appointment as Warden of the Mint on 19 March 1696. Soon to be the lover of Newton's half-niece (Catherine Barton before her marriage to Conduitt), Montagu was now senior member of the Treasury and President of the Royal Society. As warden (the representative of the Crown in the Minting process), Newton faced a number of challenges. Britain required deep financial reserves to support its military campaign against France, while the practice of 'clipping' coins of the realm had seriously degraded the value of money and the quality of coinage. Furthermore, since they contained a higher proportion of silver than the older 'hammered' coins, the new and heavier milled coins could be melted down at profit and counterfeit money made out of a mixture of clippings and copper. Early on Newton was asked for his advice on the silver question, and he argued that the melting down of coin (on account of the raw metal being worth more than the face value of the coin) was a temporary disaster that could be alleviated in the short term by allowing the circulation of paper money issued by the recently founded Bank of England. He also shared the older view that spending good money on foreign luxuries was an affront both to personal morality and to national strength.

The only long-term remedy to this was to call in all the 'old' money and dramatically increase the amount of 'new' money produced by the Mint. The 'Great Recoinage' would produce highly standardized coins with a visible edging, all manufactured by state of the art rolling mills. Although the post had previously been a sort of sinecure, Newton dedicated himself to the recoinage and – to deal with the vast amount of bullion required – the creation of temporary mints in Norwich, York, Chester, Bristol, and Exeter. Despite the extensive work of these mints, there were few silver coins left in circulation by the time Newton died in 1727.

As warden Newton was also responsible for prosecuting clippers and coiners, and recommending them for execution if the crimes warranted it. He pursued miscreants with the same intensity and indeed the same techniques he had used to prosecute corrupters of scripture. He performed a wide-ranging analysis of the art and history of clipping and coining, paid informers for information, and forwarded money to friendly witnesses so that they could dress well in court. Some jailed coiners threatened to shoot him, and in turn he showed little sympathy to criminals like William Chaloner, whose pleas for mercy in the days before he mounted the scaffold fell on deaf ears. Under Newton, the extent of clipping and coining diminished, and the numbers of individuals executed for this crime fell to zero.

By 1698, he had virtually taken over the roles ordinarily played by the Master of the Mint, at this time Thomas Neale, and he succeeded to Neale's position when he died at the end of 1699. The master was responsible for the quality of the metals in the ingots that were used to make coin, and he was responsible for laying out the income of money from the coinage.

His knowledge of chemical processes was occasionally useful in later years, especially at the so-called 'trial of the pyx', at which the quality of randomly chosen coins was tested against 'trial-piece' coins that were held by a group of goldsmiths. Occasionally too, the master was involved in procuring (and in Newton's case, designing) coins minted in different metals to celebrate royal accessions or military triumphs.

Newton lived well in London, but showed little interest in literature or the theatre. Indeed, he once told Stukeley that he ran out and away halfway through the only opera he ever attended, though one might wonder how he lasted that long. Performing as anonymously as he had over a decade earlier, he was elected as an MP in 1701 and served in the parliament that lasted until May of the following year. In May 1705 he stood again, once more with the backing of Halifax, though he suffered an ignominious defeat. The knighthood he received the previous month from Queen Anne, when she stopped at Cambridge on a visit to the races at Newmarket, was some consolation.

Opticks

Perhaps hastened by the death of Robert Hooke in March 1703, Newton was elected – by no means unanimously – as President of the Royal Society in the following November. This reignited his interest in natural philosophy for the first time in years and he used the opportunity to make his optical views available to a much wider audience than could understand the *Principia*. His *Opticks* appeared in February 1704, with 'De Quadratura' and an analysis of 'lines of the third order' tacked on at the end.

Opticks consisted mainly of old material but the fact that it was published in English, consisted largely of experiments, and avoided the abstruse mathematics of the *Principia*, made it accessible to a wide audience. It did contain a new if short book on diffraction, and in the same (third) book Newton inserted a set of 16 short 'Queries' that addressed fundamental features of his natural philosophy. Expressed as questions, these were largely couched in the *Principia*-language of 'attractions', and like the *Principia*, they did not make use of the concept of an aether. In a draft introduction to the work he argued that one should derive three or four 'general presuppositions' from a wide range of phenomena and then account for all the phenomena of the world in terms of these phenomena. Unless one began with phenomena and derived general principles from them, 'you may make a plausible system of Philosophy for getting your self a name, but your systeme will be little better then a Romance'. The account of the relationship between light and bodies in the published 'Queries' was explicitly couched in terms of micro-forces that acted at a distance, and he attacked any efforts to explain light as variations in 'motion pression or force' as 'a systeme of Hypotheses'.

As Newton wrote this, however, he was already working on a dramatic new set of seven queries, which appeared two years later in the Latin translation (*Optice*). Along with eight more that were added in the second English edition of 1717, these were arguably the most influential texts for 18th-century chemistry and natural philosophy. Here too, recalling the extraordinary analysis in 'De Gravitatione', he spoke for the first time in public about his understanding of the way God was connected with His Creation. In query 20 (query 28 in the 1717 edition) he suggested that empty space was like the 'sensorium' of God, and that God was aware of everything that took place in the universe in the same way that humans were aware of images that came into their brains. In a draft for the wide-ranging query 23 (query 31 in 1717) he noted that the Ideas of the Supreme Being 'work more powerfully upon matter than the Imagination of a mother upon an embrio'. Recalling the classical scholia, Newton also told David Gregory at this time that by His intimate presence God was the immediate cause of gravity. The claim in the drafts that space *was* God's sensorium, or was the body of God, was effectively an old Christian heresy and although this sentiment found its way into early examples of *Optice*, in later printings Newton corrected the passage to say that space was *like* the sensorium of God.

Developing the analysis of the suppressed 'Conclusio' of the *Principia* – and by extension earlier work in philosophy and alchemy – the new queries described a range of chemical phenomena that he grouped under the heading of 'active principles'. These wonderful texts brought together many of the disparate research programmes on which he had been engaged for the last four decades. In drafts for the extensive query 23, he remarked that 'the variety of motion (which we see) in the world is always decreasing', and this could only be recovered by active principles that gave rise to gravity and the numerous phenomena associated with fermentation and cohesion. These were susceptible to general rules or laws that were the 'genuine Principles of the Mechanical Philosophy': 'We meet with very little motion in the world', Newton claimed, 'besides what is (visibly) owing to these active principles, & the power of the will'. In these drafts Newton contrasted the nature of bodies that only had the passive power of inertia with the way that 'fermentation', life, and will introduced new motion into the world, fermentation being described as a 'very potent active Principle which acts upon [bodies] only when they approach one another'. 'We find in ourselves a power of moving our bodies by our thoughts', he went on, 'but the laws of this power we do not know'. Remarkably, he added that 'we cannot say that all nature is not alive'.

As soon as Newton became president, the Royal Society began to give regular payments to the instrument-maker Francis Hauksbee to perform experiments with an air-pump at their weekly meetings. From 1706 until his death in 1713 he produced extraordinary effects related to the phenomena of capillarity and electro-luminescence. Newton, who presumably advised Hauksbee on the content of many of these experiments, came to believe that the existence of this force had been demonstrated by Hauksbee's experiments and argued that electricity was a basic force operating in many other phenomena. In the General Scholium, added to the 2nd edition of the *Principia* of 1713, he announced that there was 'an exceedingly subtle but material' 'electric spirit', which was hidden in 'all gross bodies', was highly active, and emitted light.

Recalling his account of electricity in the 'Hypothesis' four decades earlier, in drafts for the eight new queries in the 1717 edition of *Opticks*, Newton attributed a multitude of short-range forces, as well as the phenomena associated with light, to this spirit. Harking back to his interest in the mind–body question and also to his work on alchemy, he even argued that the electrical spirit united 'the thinking soul and unthinking body', and could be of great use in vegetation, 'wherein three things are to be considered, generation, nutrition & preparation of nourishment'. However, in the same 'Queries' Newton reintroduced an aether that explained the relationship between light and heat. Another aether also accounted for gravitation by being composed of repelling particles, which made it highly 'elastic' – a description almost identical to that expressed in his 'Hypothesis' of 1675.

A cunning and perverse man

Although he could be sweetness personified to those who genuflected before him, Newton had what even his friends believed to be an innately suspicious temper that could erupt when his status, honour, or competence was threatened. Newton's relations with John Flamsteed, never recovered from their earlier cooling. Things came to a head in 1704 when Newton presented Flamsteed with a copy of *Opticks*, whereupon Flamsteed had his one-time assistant James Hodgson give lectures in London indicating the 'mistakes' contained within it. Anxious for Flamsteed's data to complete his lunar theory, Newton told him that he was prepared to recommend to Queen Anne's husband, Prince George, that he support the publication of a catalogue of Flamsteed's observations. As Flamsteed put it later, 'I was surprised at this proposition [having] always found him insidious, ambitious & excessively covetous of praise & impatient of contradiction'. From now on Flamsteed steeled himself against Newton's wiles, being unwilling to put himself 'wholly into his power & be at his mercy who might spoyle all that came into his hands'.

As we have seen, in the late 1690s Flamsteed already believed that Newton was overly influenced by a whole set of 'flatterers' and 'cryers up'. These he condemned as 'some few busy arrogant & self-designing people', who constantly pestered Flamsteed about the completion of the catalogue while doing everything they could to prevent it. As he suspected, Newton was showing them materials that Flamsteed had asked him to keep private, and they in turn were using this data to belittle the Astronomer Royal. Despite all this, and Newton's violent outburst against him in the winter of 1698–9, he told a correspondent in 1700 that Newton was a 'good man at the bottom but through his Naturall temper suspitious'. However, after 1704, he always saw Newton as a power-crazed despot who was actively working to 'spoil' his work.

In a move mirrored by his later treatment of Leibniz, Newton set up a committee of experts or 'referees' to oversee the production of the star catalogue towards the end of 1704. Dismissing them as weak or mere lackeys of Newton, Flamsteed believed that Newton was trying to gain all the honour for his own work, while withholding payment from the prince's fund that would have allowed Flamsteed to complete the parts of the work he wanted. After Newton was knighted in 1705 Flamsteed frequently referred to him merely as SIN, and as the years wore on he repeatedly contrasted his own 'sincere & honest' demeanour with what he called Newton's 'cunning', 'vexatious pretences', and 'disingenuous & malitious practices'.

In April 1706 Flamsteed was forced to hand over the part of the catalogue that had so far been completed, despite his protestations that it was incomplete and that it would be foolish for him to give such important work to someone else. As a precaution it was sealed by Hodgson, although Newton was upset that this seemed to reflect on his honesty. According to Flamsteed, Newton now began to accuse him of stupidity and the sabotage of his own work, while Flamsteed complained privately of Newton's increasing perversity in not paying him for his work. In March 1708 Flamsteed handed over to the referees a copy of all the observations made between 1689 and 1705. He also signed an agreement that he would hand over lunar observations and also a revised catalogue of the fixed stars, with their 'magnitudes' added. In the following years Flamsteed continued to add new observations to his star catalogue, and was left relatively free from interference from Newton. In private he frequently denounced Newton's perverse 'cunning', interspersing his comments with condemnations of Newton's work on optics and gravitation.

The brief lull in their battle ended abruptly in December 1710, when Flamsteed received an official edict from Queen Anne telling him that, in order to improve navigation, the observatory was to be overseen by a Board of Visitors – headed by the President of the Royal Society – empowered to demand at yearly intervals all of the Astronomer Royal's observations for the previous year. To make things much worse, in spring of the following year, Flamsteed heard that he was now being asked to supply some magnitudes for the constellations that had not been included in the catalogue to Newton, indicating that the latter had broken his promise and unsealed it. Alarmed by Newton's despicable actions, he was further dismayed to find that the work (the *Historia Britannica Cœlestia*) was now being printed without his input, a move that he thought 'was one of the boldest things that ever was attempted'. His fears were confirmed at the end of March 1711, when he was told that Halley was 'taking care' of his catalogue. Over the following months Flamsteed was further humiliated by being asked to correct sheets from Halley's edition, and he decided to produce his own.

Affairs came to a head at a meeting at the Royal Society's headquarters in October 1711 when Newton offered to repair the observatory's instruments, the implication being that they were the state's property and not, as Flamsteed insisted, his own. Flamsteed's delightful account indicates that Newton completely lost control, broke out in a passion & used me as I was never used before in my life: I gave no answers; but onely desired him to be calmer, moderate his passions, thankd him for the many honorable names he gave me & told him God had blest my endeavours hitherto.

According to Flamsteed the least offensive of the things Newton called him was 'puppy'; he asked Flamsteed what he had done in the nearly four decades that he had received state funding, whereupon the plucky Astronomer Royal asked Newton what he had done to earn his £500 per annum as Master of the Mint. Worse, Flamsteed mentioned that others had claimed that a passage in the *Opticks* (presumably the uncorrected remarks on God's sensorium) left Newton open to the charge of being an atheist. Along with his claim that Newton and his henchmen were robbers, this caused Newton to call him proud and insolent. Halley's edition appeared the following year, with a thinly veiled attack on Flamsteed's dilatoriness in releasing his observations. Bitter to the end, Flamsteed survived another decade, to be succeeded as Astronomer Royal by the editor of the bastard *Historia*.

Breaking Leibniz's heart

A much more substantial intellectual foe was the German Gottfried Leibniz, arguably Newton's only intellectual equal in the period. Leibniz had visited England in 1673 and 1676 and by the second visit had devised a very different version of the calculus, by now a decade old. At this stage Leibniz and Newton had a good relationship, expressed in the two letters written by Newton to Leibniz in 1676. Probably in ignorance of Newton's priority in discovering the calculus (although Collins had shown him a version of 'De Analysi' during the second visit to London) Leibniz published the rules of differentiation and integration in 1684. At the end of the seventeenth century Fatio suggested that Leibniz's calculus was both inferior to and later than Newton's, adding that it was possible that Leibniz had 'borrowed' it from Newton. In turn Leibniz wrote anonymous reviews both of the 1704 'De Quadratura' and also of 'De Analysi' (which first appeared in a collection edited by William Jones in 1711), in which he insinuated that the fluxional calculus was merely his own differential calculus in a different notation. In the following years the issue would explode into a series of bitter exchanges concerning theology, metaphysics, natural philosophy, and mathematics.

While troubled brewed with Leibniz, Newton worked with the gifted Plumian Professor of Astronomy, Roger Cotes, to recast the *Principia*. From the early 1690s Newton had worked periodically to correct his masterwork, but after they teamed up in 1709, Cotes prompted him to make more radical changes, especially to book 2. In early 1713 Newton completed the General Scholium to the *Principia*. In it he lambasted the 'hypothesis' of vortices and went on to assert that the restorative role of comets and indeed the entire ordered structure of the cosmos was proof that the world had been created by a wise and omnipotent deity. This spiritual being, he wrote, ruled over servants in a dominion as 'Lord over all'. God was present everywhere and at all times, and had a 'substantial' presence without being subject to the usual phenomena that affect bodies. There were things that could be known about God by analogy, and Newton harked back to the analysis in 'De Gravitatione' by claiming that God was 'all eye, all ear, all brain, all arm, all power to perceive, to understand, and to act'. However, this was 'in a manner not at all human . . . not all corporeal, in a manner utterly unknown to us'.

At the last moment Newton noted that discoursing about God 'does certainly belong to experimental philosophy', the expression being broadened to cover all of natural philosophy in the third and final edition of 1726. In magnifying the role of God to such an extent, Newton was as ever, cautiously expressing an aspect of his core theological beliefs. In 1713 it would still have been disastrous to be outed as an anti-trinitarian, as Whiston had been only a few years previously, though a number of divines had suspicions about the orthodoxy of the General Scholium when it was published.

In two final paragraphs he returned to the twin planks of his overall scientific project. First, he asserted that there was no need to concoct a hypothetical cause for gravity when observations and experience proved its existence. He also drew attention to 'a certain and most subtle spirit which pervades and lies hidden in all gross bodies', giving rise to the phenomena of cohesion, light, electricity, and the power we have to move our own bodies. However, he remarked, these things could not be explained in a few words, and there were insufficient experiments to determine the laws that governed them. At the other end of the work, Cotes helpfully wrote a preface in the spring in which he termed a 'miserable reptile' anyone who thought one could derive the system of the world by thought alone, or who believed God had created a cosmos whose perfect working effectively denied a role for freewill or supernatural intervention. As trouble brewed with Leibniz and his supporters, the unnamed target was clear.

Issues of priority

The so-called priority dispute had got properly under way when Leibniz responded in March 1711 to a paper by John Keill, which asserted that Newton had been the first to invent the calculus. Newton, by now having seen the 'anonymous' review of 'De Quadratura' at which Keill had taken umbrage, helped Keill draft a robust response to Leibniz's own priority claims and the latter duly replied early in 1712. Soon afterwards Newton received Leibniz's negative review of 'De Analysi', and immediately set about creating a committee of the Society to decide (as Leibniz had requested) the truth of the matter regarding the priority dispute. As in the case of Flamsteed, Newton compiled a subservient but allegedly impartial committee that was hardly likely to find in Leibniz's favour. He used his extensive forensic skills to scour his own papers and letters (including those in the collection of John Collins, which Jones had used for his edition) for evidence, presenting the committee with all they needed to reach a decision. The relevant data were collected and published under the title of the *Commercium Epistolicum D.*

Johannis Collins, et aliorum de analysi promota which appeared early in 1713.

Leibniz, excoriated throughout the text, replied anonymously through what he termed a charta volans or 'flying sheet'. He also invoked the testimony of a 'learned mathematician' (Johann Bernoulli) to the effect that Newton lacked sufficient expertise in calculus to be considered its inventor. Attack and counter-attack was launched on the pages of major European journals, and when he felt that his position was not being made sufficiently clear, Newton published his own utterly self-indulgent 'Account' of the *Commercium Epistolicum* in early 1715.

An equally important context for the dispute was the position of Leibniz as royal historiographer for the regime of Prince George in Hanover. When Queen Anne died without issue in the summer of 1714, the Hanoverian ruler became the British monarch under the Act of Settlement of 1701. Newton and his allies soon set out to convince the Hanoverians of the truth of the Newtonian philosophy. Newton arranged for optical experiments to be shown to the king's mistress, while Samuel Clarke, a chaplain to the king, began to work on the talented Princess Caroline, the wife of the Prince of Wales. However, in November 1715 Leibniz remarked to the princess that the Newtonians followed Locke in holding souls to be material and believed space was the organ of God's body by which He perceived what was going on in the cosmos. Such an accusation needed a response, and Clarke, Newton's most trusted friend in the last two decades of his life, offered himself as the man to defend Newton's cause.

Newton did not want to be drawn publicly on all these issues but the stakes could not have been higher. When a number of foreign scientists and astronomers visited London in 1715, Halley and Newton showed select visitors Newton's old and browned mathematical manuscripts to demonstrate Newton's priority in the calculus dispute. Hauksbee's able successor, Jean-Théophile

Desaguliers, also showed them Newton's crucial experiment and word got back to French philosophers that Newton's incredible doctrines on light and colour were true. In the next few years Newtonian tenets swept across the Channel: a 2nd edition of *Optice* appeared in 1719, and successive French editions appeared in the following two years.

The momentous correspondence between Clarke and Leibniz was conducted through the medium of letters to Caroline. Embracing many topics, it encapsulated all the major differences between the two camps, each caricaturing the other side in order to make their opponent's views look ridiculous or irreligious. Newton kept a close eye on Clarke's side of the dispute and, even if Newton did not draft them for him, Clarke's letters are fully consistent with his views. In an exchange of ten letters with Clarke in the year up to Leibniz's death in November 1716, Leibniz launched a number of charges, including the claim that the Newtonians made space into God's body; that God created such an imperfect world that He has to periodically intervene to fix his flawed machine; and that in holding that God acted in a way that was unconstrained by logic, the Newtonians made him into an arbitrary ruler (and by implication that the Newtonians were hostile to George I and yearned for the arbitrary rule of the son of James II). The doctrine of 'attraction' was incomprehensible, took philosophy back to the dark ages, and undid all the good work that the mechanical philosophy had established.

Clarke repeated the crude accusation that Leibniz's notion of pre-established harmony denied freewill and repeated Cotes's point that Leibniz's 'absentee landlord' God had made such a perfect machine-like Creation at the outset that He had no need to be concerned with it thereafter. Leibniz had apparently restrained God's power by suggesting that He had to obey the laws of logic, while in a similar vein, Leibniz apparently believed that from logical principles one could derive truths about the world without having to do the hard work of experimentation. For Newton and Clarke, God was omnipotent and could do things freely by the mere act of His will to achieve ends that might well be incomprehensible to mere humans (even Newton). Attraction was to be understood as a 'name' that designated an observationally based truth, much to be preferred to the obscure and overly metaphysical 'monadological' philosophy that Leibniz offered. The dispute was terminated by Leibniz's death in November 1716, by which time opinions had hardened. To Leibniz's chagrin, however, his 'pupil' Princess Caroline does seem to have moved towards the Newtonian position by the time of his death.

Chapter 10
Centaurs and other animals

In the final decade of his life, Newton continued to perform many of his administrative duties in the Royal Society and the Mint, although his health increasingly failed him. In 1725 he was advised by Catherine and John Conduitt to move to the healthier climes of Kensington – then far away from the baneful smoke of London. His intellectual energies also waned, although he devoted hours of each day to the study of prophecy, the history of the church, and chronology. A 3rd edition of the *Principia* appeared in 1726, edited by Henry Pemberton, although this added little to the 2nd.

Long finished as a creative force, Newton nevertheless remained the pre-eminent natural philosopher in Europe. For decades he had placed his disciples in top positions in major Dutch and British universities, and when no such positions were available, admirers preached the Newtonian philosophy in numerous books and lecture series. By the 1720s the Newtonian system reigned supreme, although it took until a decade after his death for his doctrines to become fully accepted in France. This was accomplished by means of the promotional skills of Voltaire, Franceso Algarotti, and Madame du Châtelet, as well as the scientific explorations to Peru and Lapland which proved the Earth to be flattened at the poles, as Newton had claimed it was.

He continued his relentless pursuit of religious truth, though he became even more cautious about reading contemporary events as fulfilments of prophecies. In a draft from the 1720s he dated the Day of Judgement to 2060 at the earliest, not least to put off those who hoped for a speedy onset of the millennium. Speculative futurology played no part in the exegetical techniques of a man who believed in accounting for prophecy in terms of historical facts. Massive drafts on early church history survive, many contemporaneous with and related to his disputes with Leibniz. These explored the earliest history of Christianity, and Newton became interested in the way that various heretical groups such as Cabbalists and Gnostics had corrupted true doctrine by means of metaphysics, 'straining the scriptures from a moral to a metaphysical sense'.

As Conduitt saw it, the most important work of his old age was a paper he entitled 'Irenicum or Ecclesiastical Polyty tending to peace'. The principles of the Christian religion were to be found in the 'express words' of Christ and the Apostles – 'not Metaphysics & Philosophy' – and these were not necessarily to be found in scripture as it now stood. All nations initially had one religion, whose basic precepts were

> to have one God, & not to alienate his worship, nor prophane his name; to abstein from murder, theft, fornication, & all injuries; not to feed on the flesh or drink the blood of a living animal, but to be mercifull even to bruit beasts; & to set up Courts of justice in all cities & societies for putting these laws in execution.

Men like Pythaogras, Socrates, and Confucius learnt this knowledge and gradually it became the moral philosophy of the heathens – 'the moral law of all nations' – even though most of them resorted to idolatry.

Idolatry was a breach of the first of what Newton took to be the great commandments – to worship and honour God. We were to give the worship due to Him to no other creature 'nor to ascribe any thing absurd or contradictious to his nature or actions lest we be found to blaspheme him or to deny him or to make a step towards atheism or irreligion'. Lust and pride – 'the inordinate desire of weomen riches & honour, or effeminacy covetousness & ambition' – were the two most egregious transgressions against the second great commandment. This was 'Humanity' – the exercise of righteousness in practice and the love of one's neighbours as one's self by treating them as one would be done by. Christianity imposed the new duty of mercy to others though – as Flamsteed observed – not everyone would have agreed that Newton ever displayed this in his own practice.

As for Christian communities, Newton claimed that all those who were baptized were members of Christ's body, or the 'church', even if they were not members of any specific church or denomination. After baptism, men were supposed to grow in grace by studying the prophecies and by comparing the Old and New Testaments, and by 'teaching one another in meekness & charity without imposing their private opinions or falling out about them'. In the Church of England, people could be received into communion by imposition of hands and they could be excommunicated if they disobeyed one of the Articles on which they were admitted to baptism, but this did not negate the membership of the larger church conferred on them by baptism. Throughout his life Newton felt able to engage in a public profession of Anglican faith while despising many of its tenets; what mattered for the chosen few like himself were their private religious beliefs.

Newton also devoted many of his last years to the study of chronology. The dating of ancient events and the euhemerist harmonizing of the histories and genealogies of different nations had attracted the attention of many of the greatest scholars in both Protestant and Catholic countries in previous centuries. Although the Old Testament was the most ancient and authentic source for ancient history, historians used various techniques to reconcile this with pagan histories that sometimes related the same events. From the late 16th century, astronomical techniques promised to aid them in pinpointing specific historical dates more precisely.

Newton's extensive researches into chronology displayed a vast knowledge of classical and Old Testament literature. In attempting to radically redate – and contract the length – of recorded history, he used dramatic new astronomical evidence based on eclipses, and adopted the extreme notion that the average length of reign of kings in history was between 18 and 20 years. Excepting Herodotus, whom he admired, he condemned the over-inflated genealogies of all the other pagan histories.

Newton was engaged in the precise dating of pre-Christian records as early as the 1680s but the vast bulk of his chronological writings date from the early 18th century, when he was Master of the Mint. An 'Abstract' of his chronology appeared first in a French translation made many years after Newton had entrusted an English version to the Venetian count Antonio Conti to pass on to Princess Caroline. The appearance of this text angered Newton immensely and gave rise to numerous refutations of its core doctrines, in particular by the great French scholars Nicolas Fréret and Etienne Souciet. Newton spent the last years of his life composing a much longer version of his writings, although this only appeared posthumously in 1728 as the *Chronology of Ancient Kingdoms Amended*.

Central to Newton's system was the dating of the Voyage of the Argonauts, at which time the astronomer Chiron the Centaur and Musaeus (master of Orpheus and himself an Argonaut) had created a 'sphere' on which the then visible constellations were drawn. By using hideously obscure evidence to locate where Chiron had placed the position of the equinoxes on the sphere, and comparing it with the value for the annual precession of the equinoxes found in the *Principia*, Newton derived a date for the expedition in the region of 936–937 BCE. Vital to his enterprise was his agreement with the Jewish historian Josephus (following Herodotus) over the identification of the Egyptian pharaoh Sesostris with Sesac, the Egyptian king who destroyed the temple *after* the death of Solomon and whose invasion of Judaea was described in 1 Kings. Sesostris (also Osiris or Bacchus) flourished in the generation before the Voyage of the Argonauts, a fact that allowed Newton to connect the dates of Egyptian history with the factual record of the Old Testament.

The birth of civilization

In the earliest times, according to Newton, there were numerous nations divided up according to the way the descendants of Noah (or Saturn) had been dispersed. Each tradition specific to a given empire called their ancestors by different names, but recounted essentially the same history. Noah's sons and their progeny lived in the Silver Age under the original seven-point Noachid law and went on to populate different parts of the world. Although the events were too early to be dated precisely, Newton waxed lyrical about life across Europe in the earliest times before the trappings of civilization appeared in the form of agriculture, beer, money, or war. In one version of a text entitled 'The Original of Monarchies' he developed his analysis of the 1680s and reasserted that the original form of worship enjoined the Ancients to practise the vestal form of worship. However, in all cases this had descended into idolatry: the Egyptians, for example, misunderstood the meaning of their hieroglyphs and their religion descended into the ludicrous beliefs of animal worship and the transmigration of souls.

Anxious to respond to the attacks that had been mounted on his system from across the Channel, Newton was working on his *Chronology* in the last 2 years of his life. Indeed, he wrote out a number of copies of various chapters of the work which only appeared after his death. The more interesting and radical elements of his great projects had disappeared, and all that remained was a filecard list of successive events. In these last weeks and months Newton apparently tried to live the ideal life he had spelt out for the good Christian, although his anger and need to crush rivals occasionally surfaced. He dispensed substantial amounts of money both to relatives and strangers and he organized the donation of Bibles. As we saw at the start of this book, both of the Conduitts recalled his great hatred of persecution and of cruelty to animals.

By the time of his death in the spring of 1727, his reputation and achievements dwarfed those of any other natural philosopher who had lived. His standing has scarcely faded in the mean time, and in terms of the extent to which anyone's scientific accomplishments surpass those of their contemporaries, Newton must be ranked above other heroes such as Darwin and Einstein. Nearly three centuries on, his private life and his 'other' academic

17. Enoch Seeman's 1726 depiction of Newton

interests continue to fascinate, while polls suggest that, worldwide, he is still regarded by most as the greatest intellect the world has seen.

Newton adopted varied approaches to problems in different areas of his work, although that is not to deny that there were connections and continuities between different strands of his intellectual research. Although it was necessarily a personal enterprise, he himself viewed his theological research as the defining aspect of his life, and the language and meaning of Scripture – along with what it said about his role in history – governed his conduct more than anything else. Respect should be paid to his intense if rather book-oriented faith, yet the astonishing courage, imagination, and originality that colour his achievements in optics, physics, and mathematics are more worthy of our admiration. As Conduitt struggled to finish his 'Life' of Newton, he came perilously close to asserting that Newton's qualities made him more than human. While he was not a divinity, there was justification in Halley's view that no human could ever get closer to the gods.

Further reading

The study of Newton's life and works has been transformed in recent years by materials made freely available online by the Newton Project (*http://www.newtonproject.ic.ac.uk*). All Newton's theological papers, and the vast bulk of his optical papers will be available by 2010, while it is envisaged that the scientific, mathematical, and administrative papers will follow in due course. The site also includes introductory assessments and articles about Newton and his work as well as a substantial number of other primary resources such as all the major published and unpublished biographical materials on Newton composed in the 18th and 19th centuries. The Chymistry of Isaac Newton project (*http://webapp1.dlib.indiana.edu/newton/index.jsp*) has already placed online many of Newton's alchemical writings and aims to make all of his work in this area available in the next few years.

As mentioned in Chapter 1, the major scholarly biographies written in the last few decades are those of Richard S. Westfall and Frank Manuel; Manuel's *Isaac Newton Historian* (Cambridge, CUP, 1963) remains the best account of Newton's chronological writings.

John Herivel's, *The Background to Newton's Principia* (Oxford, Clarendon Press, 1965) and *Unpublished Papers of Isaac Newton*, ed. A. R. and M. B. Hall (Cambridge, CUP, 1978) reproduce significant *Principia*-related drafts and revisions. Newton's optical Lucasian lectures are reproduced in vol. 1 of Alan Shapiro's projected 3-volume edition of Newton's optical papers (Cambridge, CUP, 1984–), while those without a thorough background in mathematics will be sorely tested by the magnificent edition of Newton's *Mathematical Papers* edited in 8 volumes by D. T. Whiteside (Cambridge, CUP, 1967–81). Eighteenth and nineteenth century biographical materials on Newton are now available in print, in Rebekah Higgitt, Rob Iliffe and Milo Keynes (eds), *Early Biographies of Isaac Newton, 1660–1885* (2 vols), (Pickering & Chatto, 2006).

Index

A

B

C

D

E

F

K

L

M

N

O

71. Choice Theory
72. Architecture
73. Poststructuralism
74. Postmodernism
75. Democracy
76. Empire
77. Fascism
78. Terrorism
79. Plato
80. Ethics
81. Emotion
82. Northern Ireland
83. Art Theory
84. Locke
85. Modern Ireland
86. Globalization
87. Cold War
88. The History of Astronomy
89. Schizophrenia
90. The Earth
91. Engels
92. British Politics
93. Linguistics
94. The Celts
95. Ideology
96. Prehistory
97. Political Philosophy
98. Postcolonialism
99. Atheism
100. Evolution
101. Molecules
102. Art History
103. Presocratic Philosophy
104. The Elements
105. Dada and Surrealism
106. Egyptian Myth
107. Christian Art
108. Capitalism
109. Particle Physics
110. Free Will
111. Myth
112. Ancient Egypt
113. Hieroglyphs
114. Medical Ethics
115. Kafka
116. Anarchism
117. Ancient Warfare
118. Global Warming
119. Christianity
120. Modern Art
121. Consciousness
122. Foucault
123. Spanish Civil War
124. The Marquis de Sade
125. Habermas
126. Socialism
127. Dreaming
128. Dinosaurs
129. Renaissance Art
130. Buddhist Ethics
131. Tragedy
132. Sikhism
133. The History of Time
134. Nationalism
135. The World Trade Organization
136. Design
137. The Vikings
138. Fossils
139. Journalism
140. The Crusades
141. Feminism
142. Human Evolution
143. The Dead Sea Scrolls
144. The Brain
145. Global Catastrophes
146. Contemporary Art
147. Philosophy of Law
148. The Renaissance
149. Anglicanism
150. The Roman Empire
151. Photography
152. Psychiatry
153. Existentialism
154. The First World War
155. Fundamentalism
156. Economics
157. International Migration
158. Newton